Gut Health Diet for Beginners

Gut Health
Diet *for* Beginners

Kitty Martone
Creator of HealthyGutGirl.com

A 7-Day Plan to Heal Your Gut & Boost Digestive Health

Photography by Nadine Greeff

ROCKRIDGE
PRESS

I would like to dedicate this book to those people struggling with an unidentifiable illness or imbalance. The fatigue, the pain, the helplessness, the misinformation and conflicting "expert" opinions and advice, the effort, and the cost may seem like a mountain impossible to traverse. Stay focused on your goals and keep your journey simple. You were designed to thrive.

Contents

Introduction VIII

Part Two
Recipes for a Healthy Gut

Introduction

I struggled with my health from birth, diagnosed with both spinal meningitis and pneumonia before three months of age. The super antibiotics that I went on saved my life but also nearly destroyed my immunity and later decalcified my teeth, along with the corn syrup–rich baby formula given at the time, which left me with mercury fillings and silver crowns on most of my baby teeth.

All through my early years, I battled illness and had issues with my weight. I fell victim to every cold, flu, and virus, and I suffered ear, nose, and throat issues as well as allergies. I was always nauseous, and was constantly hurting myself because I was weak. I couldn't keep my eyes open in school, and my grades suffered. By the time I reached puberty, all my symptoms compounded and others revealed themselves. As a young adult, I could barely hold down a job because of the chronic illness. When I finally reached a dead end with conventional medicine, I sought alternative help and education.

Around this time I met my husband, the best chiropractor in the world. He was the first "messenger" who came to me with the message of whole-body healing—the idea that you cannot treat your symptoms and expect to thrive, but rather you must address the root cause. This was new to me. His chiropractic method and his holistic view on food and supplements started my healing journey. Our meeting also led me to my education, which continues to this day. I became a holistic nutritionist, a master herbalist, a holistic health practitioner, and a chef, and I even spent two years working as a childbirth educator.

I also met my next two mentors, my practitioner, Dr. Robert Marshall, and Donna Gates, founder of Body Ecology and author of *The Body Ecology Diet*. This is where my specialized education began. They both spent a tremendous amount of time and attention studying what you eat and how you are digesting it. Sounds simple, but these two factors are crucial when it comes to gut health. Dr. Marshall started me on a healing program, and I began the Body Ecology Diet. Through the wisdom of these nutritional giants, I grew both personally and professionally and began to see how I, too, could contribute to this field and help educate others on the importance and value of gut health. But meanwhile, I struggled with my own health.

My exposure to so many diets and health programs as well as my work with so many clients helped me to see a common thread with diets meant to help people with chronic health concerns and gut problems. Most of these diets were too rigorous, too challenging, and often, the extreme nature of these diets would catapult people into detoxification, or what are referred to as Herxheimer reactions. This phase of adapting to a new diet can make some people worse before they get better. Most diets do not do anything to help mitigate this natural transition, which is primarily caused by the drastic shift that occurs when you cut out the sugar and toxic ingredients and flood the body with healthy minerals and probiotics from clean foods. I liken it to doing a full spring cleaning without buying any garbage bags to throw out the trash. This happened to me. And as much as we all want our houses to be thoroughly cleaned in one day, setting aside a few days to take on the whole process is far more productive than a quick fix, and the results will last a lot longer.

I personally felt much worse for almost two years before starting to feel real improvement. I remember during that time, I wished for the very type of guidance that this book offers: a more gentle and handheld transition. That phase I endured was enough to make anyone quit. And *that* is what inspired this book. I wanted to help mitigate the microorganism "die-off" reactions in people, as those reactions can be enough to make the toughest person quit. In looking for a kindler, gentler way to adjust the body, I found that the power of cooking at home and "cleaning up" your ingredients is enough to make big shifts in your energy and how you feel in as little as seven days. I also found that increasing the daily intake of vegetables through whole food vegetables and greens smoothies, as well as how you combine your food when you eat it, has a tremendous and immediate impact on the gut.

For many people, a radical, extreme diet is not necessary; even if your gut concerns seem extreme, many issues can be helped with some foundational basics that I address in this book. I have combined the same general gut health principles found in all the specialty gut health diets. These include using clean, anti-inflammatory ingredients, lowering sugar, eliminating unwanted chemicals and preservatives, and including lots of whole food prebiotics and probiotics. In fact, this diet is what I like to call "The Body Ecology Diet on training wheels." It is a great way to dip your toe into gut-healthy eating before committing to a more restrictive diet, if needed.

Are you ready to learn more? Let's get started!

PART ONE

The Gut Health Diet Solution

In part one of this book, you will learn why gut health is at the foundation of all wellness and discover some of the fascinating research in an exploration of our microbiome. You'll gain a general understanding of gut dysfunction and many of the disorders that can arise from a gut in distress. I'll introduce you to the dietary principles of this diet and offer some helpful tips on how to set up your kitchen and other useful secrets to success.

Chapter One

It Starts with Your Gut

You will one day be able to look back on your life and exclaim, "I was alive when medicine as we knew it changed!" We are literally living in a time when the entire medical model of how we view health is making a profound, monumental shift. In your lifetime, scientists went from thinking the human body was made up of and controlled by about 37 trillion human cells to realizing that not only are we made up of 10 times as many bacterial cells, but these amazing bacteria are the ones truly in charge of the human body. Science has discovered that the bulk of these microbes reside in the gut, and these little organisms make up about 85 percent of our immune system. This makes it such a fascinating, exciting field of study. Our guts rule!

How Gut Health Affects Overall Health

The research is not only "in"—it keeps coming in. Every day something new is discovered about the role that the human gut plays in nearly every single metabolic function of the body. How healthy your gut is or isn't may be at the root of every imbalance you suffer from, from depression to bloating, from cancer to toe fungus. The time to understand how to nourish your microbiome is now.

YOUR SECOND BRAIN

Science has spent most of its lifetime understanding that our central nervous system is the main computer of the human body. The central nervous system encompasses the brain and the spinal cord, with all its nerve endings that reach out to the tips of our fingers and toes and every tiny space in between. Science knew that we also had an enteric nervous system, which is a collection of nerves and neurons in the intestine that seemed to act independently of the central nervous system. This enteric nervous system has been referred to as the "second brain." However, one amazing revelation that has surfaced over the past 20 years is that the central nervous system and the enteric nervous system don't act independently at all. In fact, they act in harmony with one another, and the most amazing conductors of this system are the trillions of bacteria that reside in our guts.

YOUR MICROBIOME

About 85 percent of your microbes, or microorganisms, are beneficial, providing vital functions essential for human survival. The rest are neutral or pathogens that have the potential to cause illness. And like your fingerprints, your microbiome is unique to you. No one on the planet has the exact same microbiome as you do—not even your twin sister, if you have one. Each microbe that lives in you and on you has developed a customized way to take care of you, because, well, you are their host. It is in their best interest to make sure that you are healthy and thriving. And each organism has its own unique way of doing this.

Some bacteria or microbes help you digest your food; others help destroy foreign invaders like viruses and flus. Some bacteria are messengers, helping send important information to the brain; others are like housekeepers, helping clean and detox unwanted waste, and some even make vitamins. Knowing this information, you can see how our most important goal should be to support our microbiome; to help it do its job, which in turn supports our overall health and helps us to thrive.

Understanding Gut Dysfunction

The term "gut dysfunction" is a very general term for a vast array of imbalances in the gut. The gut itself refers to everything from your mouth to your anus. That is a lot of ground to cover. Here is some useful information to help you begin to understand the definitions and types of gut dysfunction.

IMBALANCED GUT FLORA

This is a term that many people are becoming accustomed to hearing these days, whether in the latest probiotic commercials or any recent health blog you've come across. But what causes imbalanced gut flora?

A recent study published by the US National Library of Medicine states that "at birth, the entire intestinal tract is sterile; the infant's gut is first colonized by maternal and environmental bacteria during birth and continues to be populated through feeding and other contacts. Factors known to influence colonization include gestational age, mode of delivery, diet, level of sanitation, and exposure to antibiotics." It goes on to say that some of these same factors continue to affect and change the gut flora throughout life.

DEALING WITH A DAMAGED GUT

Some other players that experts believe have a big part in negatively affecting our gut flora are certain chemicals in our food, like those found in pesticides, along with household products and even beauty products. Also cited for their detrimental effects on the gut are the consistent ingestion of refined sugars, highly processed and preserved foods, antibiotics and other medications, and two very unexpected players that nobody saw coming: electromagnetic exposure from computers, smartphones, televisions, microwaves, remote controls, and Wi-Fi devices. And *stress*.

How does this affect us? With disturbed and imbalanced gut flora, it becomes difficult to absorb the nutrients in our food. Once this imbalance begins, sometimes as soon as birth, all sorts of things can go awry. If we can't fully absorb the nutrients in our food, over time this can result in nutritional deficiency, like a lack of vitamin B, C, or D. If we can't effectively fight the "bad guys," you can see how these pathogens might start to take over and make it harder to digest food, which can compromise the gut lining and cause problems.

INTESTINAL PERMEABILITY: Also known as leaky gut syndrome, this happens when the lining of the intestine becomes thinned and microscopic fissures or cracks occur, causing particles of undigested food and bacteria to wander or leak into the bloodstream. This can cause allergic reactions in the body and lead to autoimmune conditions.

AUTOIMMUNE CONDITIONS: When the gut is permeable and bacteria and rotting food escape into the blood, the body produces an immune response, which includes inflammation and oxidation. When this happens chronically over time, it can result in issues like fibromyalgia and rheumatoid arthritis, restless leg syndrome, Hashimoto's thyroiditis, Graves' disease, and ulcerative colitis, just to name a few. There are over 100 autoimmune conditions.

Other conditions can arise as well, like:

GI DISORDERS: These include irritable bowel syndrome, constipation, diarrhea, and food allergies.

MENTAL HEALTH ISSUES: Recent science is showing that a damaged gut and compromised gut flora can be responsible for mental health issues. This research is astounding and is changing lives. The idea that healing our gut dysfunction can have a positive impact on our mental health and could potentially eliminate the need for mood-stabilizing drugs and antidepressants gives hope to many.

The Importance of a Healthy Diet

As mentioned previously, there are many factors that can contribute to gut dysfunction from childbirth, but one of the most impactful is diet. As much as refined sugar, processed and packaged foods, energy drinks, commercial dairy, alcohol, and fried foods can have a negative impact on your gut, so too can a healthy diet positively impact your gut. Certain amino acids and fatty acids as well as some minerals and vitamins can literally repair the damage that has taken place on the lining of the intestine due to erosion from inflammation and acidity. There are also numerous foods that contain beneficial bacteria and prebiotics that can help heal the stomach, as well as feed the healthy microbes you already have. There are ways you can eat that can mitigate further damage to your gut and build and support a healthy microbiome. Some experts say that healthy "upregulated" or regular digestion is the first step to being healthy. Research shows that some foods hinder healthy digestion and other foods support healthy digestion, so let's learn what those foods are.

The General Gut Health Diet

Our society seems to be obsessed with extreme diets and workouts that involve "all or nothing" types of restrictions. Yet in my experience, very few people are "all or nothing" types of people. In my 12 years of practice, I have found that no matter how excited a person is at the beginning of a diet, if they go full throttle right away, they almost always burn out within the first few weeks—me included. In addition, most diets that are geared toward balancing gut dysfunction are extremely challenging. I know because I have literally tried them all, and that's why I created this diet. And though they can all ultimately have a great positive impact on our health, I also find that symptoms of withdrawal, microorganism die-off, boredom with the program, and demanding preparation and labor can easily shift you from motivated to defeated.

Here are seven dietary rules I recommend you follow for a healthy gut:

START YOUR MORNING WITH WARM LEMON WATER. A quart of warm water with half a lemon squeezed into it is the perfect way to rehydrate after a long night of sleep with no water, and it also stimulates digestion.

CHEW YOUR FOOD THOROUGHLY. Most people eat on the go or under stressful situations; sometimes we are distracted and just don't chew our food. This immediately sets us up for challenged digestion.

EAT FERMENTED FOODS. Fermented foods are already digested by the beneficial bacteria that helped to ferment them, so they are very easy on digestion. They are also almost 100 times more nutritious than if they were not fermented (see Getting to Know Fermented Foods, page 6).

EAT PLENTY OF FIBER AND PREBIOTIC FOODS. These are food for our microbiome. Science is learning that it isn't so much that we need more beneficial bacteria in our gut, but that we need to feed the bacteria we already have.

AVOID OVERCOOKING YOUR FOOD. Natural foods contain beneficial enzymes that help you to digest them; however, when you overcook food, you destroy these enzymes, making it harder for your body to break down these foods and adding to sluggish digestion.

ADHERE STRICTLY TO THE "FOODS TO AVOID" LIST. The problem with the foods on this list is simply that they are not in their natural state, and our gut has a more difficult time breaking them down, thus resulting in sluggish digestion and an inflamed gut. Our guts like natural things.

FOLLOW BASIC FOOD COMBINING RULES. Simple meals digest the best. Alkaline foods require different digestive enzymes than acidic foods to be broken down in the gut. Fruit requires a totally different enzyme to be broken down than beef, and one actually neutralizes the other. As a result of combining the wrong foods, the whole digestive process slows down, bloating occurs, and putrefaction begins.

Following are some basic food combining rules. There's no need to obsess about these; they are only mentioned to help you identify why certain food combinations might make you feel better or worse after meals.

- Eat fruit alone, especially melon.
- Pair protein with non-starchy vegetables.
- Pair starches with healthy fats and vegetables.
- Leafy greens and non-starchy vegetables go with everything.
- Drink water away from meals. Most beverages dilute stomach acid and slow digestion.
- Spices and herbs are neutral.

GETTING TO KNOW FERMENTED FOODS

If you are unfamiliar with fermented foods, just know that if you like pickles, you will probably love fermented vegetables. But for those who aren't fans of the sour flavor, it can be an acquired taste. Ferments are fizzy, sour, tangy, and usually pungent.

Fermented vegetables are vibrant in color, crunchy, and usually salty. Fermented beverages are usually vinegary, sour, and tangy and really good ones are fizzy or lightly carbonated. The best part of ferments is that they are LOADED with nutrients and beneficial bacteria and so easy on the digestion.

An interesting thing to consider is this: When a person has severe gut imbalance, the microbes that are running the show crave simple sugars. That means that you will crave foods that are sugary or break down into sugar easily, like bread. When you begin to establish a healthy microbiome, the microbes that are running the show prefer a more diverse variety of flavors, like bitter and sour as well as sweet and savory. And with that, your palate for fermented foods grows. And all of a sudden, something like a strawberry is so incredibly sweet, and something like a piece of candy is practically inedible because it's overly sweet. I see it happen over and over with clients who are initially afraid to touch ferments, and as soon as a month later, they are eating them as an entire meal!

FOOD SENSITIVITIES, GI DISTRESS, AND OTHER CONSIDERATIONS

This diet is a general gut health diet, and there may be a few foods recommended on the list that you do not do well with. Perhaps you have an allergy to a food mentioned, or perhaps you experience discomfort when you consume a food. You should never continue to eat a food if it gives you any type of unwanted reaction. Keep in mind, if you've never had fermented foods, you may find that you experience some bloating when beginning to eat these foods. This is normal, but it should not be intolerable or painful. Some recipes have modifications offered for the Big 8 allergens (wheat, peanuts, tree nuts, fish, shellfish, dairy, soy, and eggs); however, these modifications will not be reflected in the weekly shopping list, so if you know you have these allergies, be sure to modify your list before shopping.

WHAT ABOUT PROBIOTIC SUPPLEMENTS?

Shocking research results are out, and they say that most probiotic supplements don't make it past the stomach acid. I personally use spore-based probiotics. My favorite is Just Thrive. Spore-based probiotics do make it past the stomach acid and in fact don't start to flourish until they arrive in the large intestine, where they belong. You will get a tremendous amount of beneficial bacteria from the diet in this book, but even more important is that we feed the ones we already have with a diet high in prebiotics.

You may ask, can one get too many probiotics? The answer is yes, particularly in supplement form. I do not encourage taking over-the-counter probiotics at will. The health of our microbiomes rely on balance, and too much of a good thing isn't always good. There are disorders like small intestinal bacterial overgrowth, which is literally an overgrowth of bacteria, both good and bad, growing where they shouldn't be, in the small intestine, that can be further aggravated by overconsumption of probiotics. It's worth the money and effort to look into spore-based probiotics. (See Resources for supplement recommendations, page 143.)

FOODS TO ENJOY

Foods that do not tax digestion, and some that even help heal the gut:

ESSENTIAL FATS AND OILS

Try and keep your refined oils away from sunlight and heat.

Avocado oil

Coconut oil

Ghee

Grass-fed raw or cultured butter (such as Kerrygold)

Olive oil (cold pressed)

Pumpkin seed oil

Sesame oil

DAIRY

Cashew yogurt

Coconut yogurt

Goat cheese

Goat kefir

Kefir (without added sugar)

Nut milks (especially homemade)

FRUITS

Blueberries

Cranberries

Goji berries

Golden berries

Green apples

Lemons

Limes

Mulberries

Raspberries

Strawberries

VEGETABLES

Arugula

*Asparagus

Avocado

Bamboo shoots

Beets

Broccoli

Broccoli sprouts

Brussels sprouts

Cabbage

Carrots

Cauliflower

Celery

Chives

Cucumber

Daikon

*Dandelion greens

Fermented vegetables

*Garlic

Green beans

*Jerusalem artichoke

Kale

*Leeks

Lettuce (all varieties)

Miso, miso paste, and natto

Mushrooms

*Onion

Parsnips

Peas

*Potatoes

Radishes

Sea vegetables (all varieties)

Shallots

Spinach

Squash

Turnips

Yams

Zucchini

* *These vegetables are high in prebiotic fiber, which means that your microbiome needs them to thrive—they are food for your beneficial bacteria.*

GRAINS AND GRAIN-LIKE SEEDS

Amaranth

Brown rice and brown rice pasta

Buckwheat

Millet

Quinoa and quinoa pasta

Spelt

Tapioca

MEATS

Bone broths

Organic/grass-fed beef

Organic/grass-fed bison

Organic/grass-fed lamb

Wild game

POULTRY AND EGGS (FREE-RANGE WHEN POSSIBLE)

Chicken

Duck

Eggs (large)

Quail

Turkey

SEAFOOD (LOWEST IN MERCURY)

Anchovies

Crab

Flounder

Haddock

Mullet

Oysters

Pollock

Salmon (wild-caught)

Sardines

Scallops

Shrimp (wild-caught)

Squid

Tilapia

Trout

BEANS AND NUTS

Always limit nut intake to one handful per day.

Almonds (raw, soaked)

Brazil nuts

Cashews (raw, soaked)

Lentils

Pecans

Pistachios

Walnuts

SEEDS

Always raw; you can lightly toast them yourself.

Pine nut

Pumpkin

Sesame

Sunflower

SPICES AND CONDIMENTS

Allspice

Apple cider vinegar, unpasteurized (I use Bragg's)

Basil, parsley, and other leafy green herbs

Cacao (raw, not cocoa)

Cardamom

Cayenne

Chili flakes

Chipotle

Cinnamon

Coconut aminos

Curry powders

Dill

Garlic powder

Ginger

Himalayan salt

Lacanto

Mint

Nutmeg

Nutritional yeast

Oregano

Rosemary

Sea salt (air dried)

Stevia

Thyme

Vanilla extract (and other baking extracts)

Yacon

Probiotic and Prebiotic Foods

It's helpful to know which foods provide probiotics (healthy bacteria) and prebiotics (food for healthy bacteria). Consider these foods in your gut health diet for their special benefits.

PROBIOTIC FOODS

Fresh herbs and vegetables (all vegetables contain high amounts of beneficial bacteria)

Homemade yogurts

Kefir

Kimchi and other fermented vegetables

Kombucha

Miso

Natto (fermented soy beans)

Raw apple cider vinegar (organic)

PREBIOTICS

Avocado

Carrots

Coconut meat

Dandelion greens and leafy green vegetables

Jicama, Jerusalem artichoke, and chicory root

Peas

Potato skins

Raw garlic, onion, leeks, chives, and scallions

FOODS TO EAT IN MODERATION

When eaten in abundance, these can hinder healing, create inflammation, and impair digestion:

FATS AND REFINED OILS
Limit to 1 teaspoon per day.
Commercially processed butter

DAIRY
Limit to ½ cup 3x per week.
Cheese (hard, aged more than 1 year)
Greek yogurt (no sugar added)
Plain yogurt
Raw milk

FRUITS
Limit to 3 servings per week.
Apricots
Bananas (½ banana per day)
Dates
Grapes
Guava
Mango
Melon
Nectarines
Papaya
Peaches
Pears
Pineapple
Plums
Red apples
Note: All juicing should be done in moderation and only if there is no digestive distress following consumption.

VEGETABLES
Limit to 2 servings per week.
Eggplant
Peppers
Russet potatoes
Tomatoes

GRAINS
Limit to 2 servings per week, only if no digestive distress is experienced after consumption.
Blue corn tortilla chips
Corn tortillas
Oats (non-GMO, gluten-free)
Popping corn (non-GMO)

MEAT, COMMERCIALLY RAISED
Limit to 1 serving per week.
Beef
Lamb
Pork
Veal

SEAFOOD
Limit to 1 serving per week.
Bluefish
Grouper
Halibut
Sablefish
Sea bass
Spanish mackerel
Tuna (canned, 3 cans per week)
Tuna (fresh, except skipjack)

BEANS AND NUTS
Limit to 1 cup per week.
Beans, home-cooked, presoaked
Chickpeas
Macadamia nuts

SEEDS
Limit to 2 tablespoons 2x per week.
Chia
Flaxseed
Hemp

SPICES, CONDIMENTS, AND OTHER
Limit to 1 serving 3x per week.
Caffeinated beverages (ideally, the goal is to eliminate caffeine)
Raw honey
Tamari / gluten-free soy sauce

FOODS TO AVOID

Foods that can hinder healing, create inflammation, and impair digestion:

FATS AND REFINED OILS
"Butter" spreads
Canola oil
Corn oil
Deep-fried fats
Hydrogenated fats
Margarine
Oils that have been heated to smoking point
Rapeseed oil
Safflower oil
Soy oil
Trans fats

DAIRY
Buttermilk
Condensed milk
Cream
Flavored yogurts
Half-and-half
Ice cream
Milk
Non-aged cheeses
Nondairy creamers
Sour cream
Heavy (whipping) cream

FRUIT
N/A

VEGETABLES
N/A

GRAINS
Barley
Bulgur
Cereals (packaged)
Chips
Cookies
Crackers
Oats
Pastas
Wheat
White rice

MEATS
Bacon (commercially packaged)
Cured meats
Dried meats
Hot dogs
Lunch meats
Processed meats
Sausages

SEAFOOD (HIGHEST IN MERCURY)
King mackerel
Marlin
Orange roughy
Shark
Swordfish
Tilefish

BEANS AND NUTS
All pre-roasted and salted nuts
Beans (commercially canned)
Peanut butter
Peanuts

SEEDS
All pre-roasted, salted, and packaged seeds

SPICES, CONDIMENTS AND OTHER
Agave
Alcohol
Artificial sweeteners like Splenda and Sweet 'N Low
Commercial dressings
Commercial iodized table salt like Morton's
Commercial ketchups
Commercial sauces
Diet drinks
Energy drinks
Fried foods, as well as foods that have been overcooked, charred, and blackened
Heated commercial honey
High-fructose corn syrup
MSG (monosodium glutamate)
Pastries
Refined processed sugar (raw, brown, coconut, palm, etc.)
Soda
Soy sauce (except gluten-free)
Store-bought products with added sugar

Chapter Two

Five Steps to Better Gut Health

Over my many years of helping clients and myself through diets and health programs, the most useful and important tool I found was goal setting and organization. It is very easy, when taking on a new regimen, to become overwhelmed with preparation and the daunting concept of making it to the end! It can be enough to throw in the towel. I highly encourage this five-step process for success.

Step One: Set Up Your Kitchen

A good kitchen setup can make the difference between success and failure—you need the right tools. Here is what you'll need for your kitchen setup.

MUST HAVE

HIGH-POWERED BLENDER: You will likely use this every day for multiple purposes. There are many brands of high-powered blenders ranging in price.

FOOD STORAGE CONTAINERS OF VARIOUS SIZES: I prefer glass, and Costco always has great specials on boxes of them. These are important for premaking foods, freezing, and preparing on-the-go meals.

4 OR 5 QUART-SIZE MASON JARS WITH LIDS: Perfect for on-the-go smoothies, shakes, and fermenting vegetables and milk.

WELL-SHARPENED CHEF'S KNIFE: I encourage spending a little more money on this item because it's the right tool for almost every job in the kitchen.

WELL-SHARPENED KITCHEN SCISSORS: Scissors are the most underutilized utensil in the kitchen. I use scissors for everything from cutting salad into a bowl to cubing meat in a recipe.

2 CUTTING BOARDS: Use one just for meats. I use the plastic ones because they are easy to keep sterile and don't take up much storage space.

LARGE SOUP OR STOCK POT: Soups are a gut's best friend, which makes a soup pot your best friend. Mine is a stainless steel 8-quart pot.

BAKING DISHES: I love my Pyrex glass baking dishes with plastic lids because after you cook with them, you can pop the lid on the leftovers and store them in the same dish you baked in. I have the set of three, but the size I use most often is the 15 x 10. And Target always has them on sale! For the recipes in this book, use the 15 x 10 unless instructed otherwise.

BAKING SHEET: I have two, a three-quarter sheet pan (most common) or 21 x 15, which you'll need for roasting vegetables, and the other is a half sheet, or 18 x 13. I use it less often, but it's good to have on hand.

SKILLETS: I have 2 stainless steel skillets, a 12-inch and an 8-inch. The larger is used more often for almost everything, whereas the smaller one is more for eggs or a quick stir-fry. If you already have a cast iron skillet, then you can always use that. But if you've never used a cast iron skillet before, now is not the time to start. Your 7-day journey on this diet will be plenty to occupy your time, and a new cast iron skillet requires some extra care.

SAUCEPANS: I have a set of 2 saucepans, a 3-quart and a 4-quart for oatmeals, rice, and quinoa, both stainless steel.

NICE TO HAVE

CROCK-POT®/INSTANT POT® OR PRESSURE COOKER: Equipment such as slow cookers and pressure cookers (including popular brands Crock-Pot and Instant Pot) reduce your hands-on cooking time, and who can't use more time? Many of the recipes in this book can be modified for use with one of these kitchen wonders.

MINI TRAVELING BLENDER (LIKE A MAGIC BULLET): If you are always on the go or have a kitchen at your office or dorm room, this is a great, compact way to stick to making smoothies and soups when you aren't at home.

HEATING PAD WITH TEMPERATURE SETTING: I find a heating pad to be super helpful when fermenting. One of the big mistakes of fermenting is letting the temperature drop too low. A heating pad ensures good ferments every time.

STAINLESS STEEL COOKWARE: So much research coming out shows that the coatings in nonstick cookware contain chemicals called PFCs and PFAs, which the Environmental Protection Agency has declared possible carcinogens. These compounds can peel off into your food, and at high temperatures they can leech into your food. A study conducted by the Centers for Disease Control and Prevention (CDC) discovered that roughly 98 percent of Americans now have traces of PFAs or PFCs in their bodies.

MESH STRAINER, NUT MILK BAG, OR CHEESECLOTH: You'll need one of these if you want to make your own nut milks and some condiments, like ketchup.

FOOD PROCESSOR: Not imperative, but it sure does take some of the labor out of shredding vegetables when you're making 6 quarts of fermented yumminess!

Step Two: Organize Your Pantry

Before you decide to throw out all your "foods to avoid," let's just find a place to keep them on hold until we know what you are tolerating. Here is a foolproof way to organize everything:

BASICS WORTH STOCKING: You can easily reference the meal plan's Foods to Enjoy (page 8) and Shopping List (page 29) to help you with this section. Favorite go-to spices like garlic salt and cinnamon, sea salt, oils, brown rice, gluten-free oats and grains, and unpasteurized apple cider vinegar would count among these items, depending on what you like to make.

ITEMS TO SET ASIDE: These might include things like peanut butter, canola oil and other vegetable oils, all-purpose flour, sugar, canned foods, packaged foods like crackers, and premade foods like ramen, regular pasta, white rice, commercial

FAVORITE BRANDS

I personally love anything that has the label "Organic" and "Non-GMO," and these days most conventional markets are beginning to carry more of these items. Even Walmart and Costco are starting to carry more organic produce. If your local market is not carrying much, it is never a bad idea to suggest to your grocer that they begin stocking organic products. Research is showing that many GMOs (genetically modified organisms) can damage gut flora, and the combination of pesticides and GMO organisms are being implicated in the cause of leaky gut. To help you stay clean with organic, non-GMO products, here are some of my favorite brands that you can find just about anywhere:

- Seeds of Change makes a packaged precooked wild rice with a few different flavors like Organic Quinoa & Brown Rice with Garlic.
- Tinkyada has various types of gluten-free pasta that are just delicious.
- Lundberg has a whole selection of gluten-free rice and grain products that I love, like brown rice cakes.
- Newman's Own brand of just about *anything*. Who isn't a fan of Paul Newman? The company's large line of organic, non-GMO products offers a lot to choose from. I love the organic pasta sauce, olive oil, and salad dressings, although do read labels; the ingredients aren't all ideal.

sauces like barbecue and hot sauce, and dressings that contain sugar and preservatives.

One of the sneakiest and hard-to-avoid issues with cleaning out your pantry is that of hidden ingredients that are in almost every premade sauce, dressing, mixed spice, and flavoring. Many hidden ingredients are very harmful to our microbiome, even when consumed in small quantities.

Here are a few hidden culprits: MSG, "natural flavorings," yeast extract, corn syrup, corn sugar, artificial flavors or the word "flavoring," artificial coloring or dyes, hydrogenated/fractionated oils, potassium/sodium benzoate, aspartame, and BHT/BHA.

Step Three: Plan and Prep Your Meals

This is by far the most important factor in achieving success: food preparation. Prepping not only "forces" you to premake and plan your foods for several days in advance, but it also keeps you from waiting until the last minute when you are so hungry you have nothing to rely on but that frozen pizza you hid in the back of your freezer months ago! This step will keep you from making mistakes that your gut will likely regret.

CUSTOMIZING MEAL PLANS

For people who are on the go, it's very helpful to make large portions or double up on portions for your week. I like to prepare a large pot of soup and then store it in individual containers in the freezer. You can take one to thaw out each night and warm up in the morning and it's ready to go. Or make a large portion of quinoa, and you'll have it handy to warm up and throw together with some quick sautéed vegetables and perhaps a piece of precooked chicken from the fridge.

Keeping a small blender like a Magic Bullet at work along with some simple smoothie ingredients can keep you from giving in to office pizza and donut days. The number one enemy of adherence to a diet is an empty stomach!

SMART SHOPPING

Now that you have your shopping list in hand and modified to accommodate any other dietary restrictions, consider shopping at grocers like Costco, where you can stock up on bulk items like sea salt, brown rice, large packs of chicken, and other staples.

ELIMINATION DIET SYMPTOM TRACKER

Use this page to track symptoms each day during the elimination diet. If you repeat the diet after the first week, continue to track symptoms, which will be especially encouraging as they disappear.

DAY	SYMPTOMS
1	
2	
3	
4	
5	
6	
7	

ELIMINATION DIET SYMPTOM TRACKER

Use this page to track symptoms each day during the elimination diet. If you repeat the diet after the first week, continue to track symptoms, which will be especially encouraging as they disappear.

DAY	SYMPTOMS
1	_____
2	_____
3	_____
4	_____
5	_____
6	_____
7	_____

If you don't already shop at your local farmers' markets, give it a try; oftentimes you can find less expensive produce than at your market, and certainly the freshest. And it's a lovely shopping experience, to boot.

I also like to plan my shopping days in advance. I recommend this so you aren't stopping at the store for a random ingredient and leaving with $75 worth of unnecessary goods. Pick two days; I prefer Mondays and Thursdays, and if you have the space in your kitchen for extra goods, I suggest picking just one day per week for shopping. Try picking a time that you know will not be busy and chaotic, and if you must visit multiple stores, plan those visits on the same day so you aren't overwhelmed with multiple days of shopping.

EFFICIENT PREPPING

As you get started, efficient prepping is probably the most daunting of all tasks, but it becomes habit quickly. Here are some key secrets to make prepping painless and set you up for success.

GET IN A GROOVE. Schedule food prep days into your calendar in advance; don't wait until the last minute.

CLEAR THE PATH. Give your fridge a good cleaning; throw out old stuff and organize and make room for new.

STOCK UP ON STORAGE. Make sure you have a good variety of storage containers to store your prepped food.

ORGANIZE BY DAY. Use individual baggies marked with the day of the week for snack portions.

KEEP LISTS HANDY. Magnet your meal plan and shopping list to your fridge or someplace visible; also keep one copy in the notes on your smartphone.

Step Four: Do the Diet

Now you've got your ducks in a row, as they say, and you are unstoppable. You have organized and shopped and prepared, and now the only thing left is to succeed. And eat! You've heard of long-term goals and short-term goals, but have you heard of mini goals? Mini goals are a great way to celebrate little successes—because I know what it's like to start a diet and literally fall off the wagon one day into it! I create these little goals on my calendar, like "48-hour goals," and I put a little star when I have completed 48 hours. Perhaps you can mark down on your calendar an achievement like "3 days without refined sugar," and when that goal is achieved, congratulate yourself and understand that these little mini steps are all adding up to a better quality of life for you.

The symptom tracker on page 19 is a very useful tool and one of the most important parts of your progress. It is important to know what foods you may have sensitivities to and which foods you do not. This will be useful information for you to have moving forward.

Step Five: Add Healthy Habits

Healthy lifestyle habits are an important part of any diet's success. Learn more about how stress management, exercise, sleep, and even a "dirty" lifestyle can help your gut health!

STRESS MANAGEMENT

You can't exercise your way out of a bad diet, and the opposite is also true. Both a sedentary lifestyle and unmanageable stress levels can easily ruin all your good dietary hard work.

Here are some ways to help manage your stress and increase activity:

1. Adopt a solid breath work practice—nothing is more powerful for stress reduction than learning to breathe properly.

2. Read up on organizational tips, and reduce clutter in your home and work environment.

3. Make time for low-impact exercise each day. Try a brisk walk while listening to a good book or some low-impact yoga. Good habits can start as simply as choosing to take the stairs instead of the elevator.

4. Carve out time for a de-stressing Epsom salt bath a few times a week.

5. Turn off your Wi-Fi at night or when you are not using it. Unplugging from the world can help manage some of your stress, according to an article in *Scientific American* called, "Mind Control by Cell Phone."

GOOD SLEEP

Sleep is more than sleep; it's the most precious time that our bodies use to detox, heal, and restore. Here are a few tips to ensure a better night's rest:

1. Don't eat large meals after 6:30 p.m. Stick to salads and soups.

2. Practice good sleep hygiene. Turn off Wi-Fi devices and televisions at least 1 hour before bed.

3. Keep a technology-free bedroom; that means no TV, computers, smart-phones, or digital clocks. Use an old-fashioned alarm clock if you need a wake-up call.

4. Understand that the bad habit of being a "night owl" can be trained away, because the human body is programmed to go to sleep shortly after the sun goes down.

5. Try adding an air purifier to your room. The white noise from the purifier is soothing, and it will allow you to breathe fresh air while you sleep.

A "DIRTY" LIFESTYLE

Dirty isn't always bad. A 2013 study showed that the microbiomes of people who lived with dogs were far more diverse than those who did not. Other research is showing that antibacterial hand soaps and gels are killing off beneficial bacteria that are there to help us fight the bad guys.

You see, our microbiome refers not only to the microbes in us, but also to the suit of armor we carry on us, made up of trillions of bacteria that are there to help protect us from pathogens. Exposing yourself to a diverse abundance of microbes in nature will help strengthen and add to this outer microbiome.

Many studies have revealed that dog owners and families that live on farms with farm animals and work the land have more diverse microbiomes than those that do not. They also have fewer occurrences of asthma, skin issues, and gut-related problems.

Here are a few tips to help you avoid killing off your precious helpful bacteria:

1. Avoid using antibacterial hand soaps and gels.

2. Before agreeing to an antibiotic, ask your doctor for a culture test. Make sure you are getting the right medication for the job. A Harvard University study showed that 73 percent of acute bronchitis visits ended in a prescription for

antibiotics, when the rate should be at almost zero percent because acute bronchitis is usually viral. In fact, antibiotic resistance is one of the biggest threats to global health, food security, and development, according to a World Health Organization report.

3. Spend time gardening and walking barefoot in the grass.

4. Reduce exposure to harsh and harmful beauty and cleaning products.

5. Make your own fermented vegetables instead of buying them. When you mix your own microbes into your ferments by touching them with your hands, they become customized to what your body needs!

6. Rescue a dog, and science shows he will help rescue your microbiome. Not only will a dog get you out in nature more, but he will happily share his gut diversity with you. If you are wondering, the study did not show as much benefit from other indoor pets including cats.

Garlic and Herb Roasted Lamb Chops with Tzatziki Sauce, page 99

Chapter Three

Your Seven-Day Diet to a Healthy Gut

In this chapter, you'll learn what to expect from this diet as you begin this exciting seven-day journey. You will learn how to set realistic, achievable expectations for yourself and become aware of the potential challenges, cravings, and symptoms and side effects that can arise as your gut restores itself. You will also be introduced to some very useful guidance, including a meal plan, shopping list, and helpful preparation guide—everything you need for a successful start.

What to Expect

As with all diets, the goal is change. We hope change is all for the better, but it can be a very natural occurrence to have some temporary side effects as well. As they say, you've got to crack some eggs to make an omelet. Here is a brief description of what you can possibly expect:

- **SOME COMMON SYMPTOMS ON THIS DIET ARE RELATED TO THE POSSIBLE DIGESTIVE RESPONSE TO YOUR INCREASE IN PROBIOTIC FOODS, PREBIOTICS, AND GREEN VEGETABLES.** They can range from an increase in bowel movements to constipation, bloating, and gas. These symptoms should not be intolerable or cause pain. You may also notice increased energy, better sleep, improved mood, and better breath. Note: If fermented foods cause you discomfort, scale back on your intake to even a teaspoon per day. If the pain persists, take a break for 7 days, until after the end of the diet, then reintroduce in very small amounts. Don't give up.
- **DON'T GET DISCOURAGED IF YOU EXPERIENCE NO SYMPTOMS.** We want there to be as little drama as possible during this transition. Sometimes no news is good news—if you are following the diet, changes are occurring, whether you feel them or not.
- **IT IS NORMAL FOR THE BODY TO CRAVE DIFFERENT FOODS WHEN YOU ARE TRANSITIONING ONTO A DIET SUCH AS THIS.** The drastic reduction in sugar will begin to starve the bad bacteria in the gut, which will generate temporary sugar cravings. With this diet, I do find that hunger cravings are not as big of an issue as with some of the other more restrictive diets. This diet includes a plethora of amazing foods that you can eat in abundance. A perfect go-to snack when feeling hungry is one of the several smoothie recipes chapter 5 (page 41) of this book offers. Don't be afraid to experiment with your own smoothie ideas as well. They never get boring.

FOOD AND CRAVINGS: BE GENTLE WITH YOURSELF

Question: What is one of the first things that brought us comfort as a crying infant?

Answer: FOOD.

We have comforted ourselves with food since day one, quite literally. There is no arguing that there is an emotional component to eating for almost every single one of us. This should be reason enough to go easy on yourself when you begin this diet and suddenly find that you cheated or had to quit and start over. When we are under stress and feeling down, we tend to crave those foods that bring us comfort, even if those foods happen to be on the Foods to Avoid list (page 11).

Don't beat yourself up! The guilt you punish yourself with is often far worse than the consumption of the food in question. Do you remember your "mini goals" that I mentioned earlier (page 20)? If you can look forward to a daily goal of something as simple as only having one cup of coffee today instead of three, or not indulging in that dessert you usually have before bedtime, that is a mini reason to celebrate. Celebration doesn't have to be reserved for monumental life changes. It is the culmination of all the little steps we take that make up our journeys. So get into the habit of celebrating your small victories!

THE MEAL PLAN

This is an example of a meal plan you can follow exactly, or you can get creative with the recipes in this book and create your own.

	BREAKFAST	SNACK	LUNCH	SNACK	DINNER
M	Monkey Madness Smoothie (page 56)	Brown rice cake with almond butter and pomegranate seeds	Arugula and Warm Squash Salad with Pomegranate Seeds and Feta (page 68)	Chocolate Chocolate Brownie (page 127)	Steamed Cod and Ginger over Wild Rice (page 80)
T	Spaghetti Squash Hash Browns and Soft-boiled Egg (page 49)	½ cup Gluten-Free Granola Clusters (page 121)	Chicken-Avocado-Lime Soup (page 107)	Berry 'n' Greens Smoothie (page 55)	Lettuce Cup Veggie Tacos (page 84)
W	Alkalizing Green Smoothie (page 54)	Green apple slices with cinnamon	Greek Quinoa Tabbouleh (page 70) with cold chicken breast	½ avocado with sea salt and lemon	Baked Salmon in a Foil Pouch (page 78)
T	Soft Scramble with Pink Cabbage Kimchi (page 43)	Papaya-Banana Smoothie with Kefir (page 57)	Leftover salmon with Baby Red Potato Salad (page 72)	½ cup leftover quinoa tabbouleh	Brown Rice Pasta with Arugula, Feta, and Pine Nuts (page 90)
F	Alkalizing Green Smoothie (page 54)	Cup of mixed berries and cinnamon	Broccoli-Fennel Soup (page 86)	½ cup raw mixed nuts (Brazil, almonds, cashews)	Simple Healthy Meat Loaf (page 94) and side mixed green salad w/vinaigrette
S	Cucumber-Avocado Toast with Aleppo Pepper (page 42)	2 Coconut Macaroon Bombs (page 124)	Leftover meat loaf and salad	Guacamole with Pink Cabbage Kimchi (page 119) with blue corn chips	Curried Chicken Legs with Carrots (page 108)
S	Spaghetti Squash Hash Browns and Soft-boiled Egg (page 49)	Carrot and celery sticks and tahini dip	Fish Chowder (page 81)	1 cup leftover vegetables from Saturday night's dinner	Garlic and Herb Roasted Lamb Chops with Tzatziki Sauce (page 99)

The Shopping List

Double-check all your labels to make sure there are no hidden or unnecessary ingredients, like sugar.

CANNED AND BOTTLED ITEMS

- Coconut milk (unsweetened)
- Coconut or avocado oil (other options include pumpkin, raw or toasted sesame, and olive)
- Gluten-free soy sauce, also called tamari
- Unpasteurized apple cider vinegar (I use Bragg's)

DAIRY, EGGS, POULTRY, AND FISH

- Chicken breasts, legs, and thighs
- Cod fillets
- Eggs (free-range)
- Feta cheese
- Grass-fed, cultured, or clarified butter (ghee) (I use Kerrygold)
- Organic, unpasteurized cow or goat milk or kefir
- Plain Greek yogurt
- Salmon fillets

MEAT

- Grass-fed ground beef
- Lamb chops

PANTRY ITEMS [ALL NUTS AND SEEDS SHOULD BE RAW]

- Almonds
- Almond butter (raw is ideal)
- Almond flour
- Brazil nuts
- Brown rice (Lundberg is my favorite)
- Cacao powder (raw, unsweetened, not cocoa powder, and nothing extra added)
- Cinnamon (ground)
- Curry powder (I like yellow curry powder)
- Dijon mustard
- Garlic or garlic salt (granulated)
- Honey (raw)
- Medjool dates
- Pumpkin seeds
- Pure maple syrup (make sure this is not high-fructose corn syrup)
- Quinoa
- Sea salt or pink salt (make sure this is simply salt with no anticaking agents or other ingredients)

- Sunflower seeds
- Tomato paste
- Unsweetened dark chocolate chips
- Vanilla extract
- Walnuts
- Wild rice (Lundberg is my favorite)

PRODUCE

- Arugula
- Asparagus
- Avocado
- Baby red potatoes
- Bananas
- Blueberries
- Broccoli, 2 medium heads
- Brussels sprouts
- Butternut squash
- Cabbage, 2 medium-size green, 1 medium-size purple
- Carrots
- Cauliflower, 1 medium head
- Cucumbers
- Fennel
- Garlic
- Garnet yams or sweet potatoes
- Ginger
- Granny Smith apples, 3 or 4
- Lemons
- Limes
- Mixed greens
- Papaya (non-GMO)
- Pomegranate seeds (You can usually find these loose in a container but a whole fruit will do if you are up for the messy task of seeding it—wear gloves!)
- Onions
- Parsnips
- Red, yellow bell peppers
- Rosemary
- Romaine lettuce (Bagged romaine hearts are easiest)
- Russet potatoes
- Spinach
- Spaghetti squash
- Strawberries, 1 pint

OTHER

- Blue corn chips (Trader Joe's has the best organic non-GMO variety)
- Brown rice cakes (Lundberg red quinoa and brown rice are my favorite)
- Sesame bread (Ezekiel 4:9 brand Sesame Sprouted Whole Grain Bread is my favorite)
- Gluten-free oats
- Gluten-free pasta
- Mixed frozen berries
- Tahini dip (Trader Joe's is my favorite)

Prep Guide

Good luck is simply when preparation meets opportunity. Here's your chance to have great luck with this diet.

WASH AND CUT

- Berries
- Carrot and celery sticks
- Root vegetables

COOK AND STORE

- Chocolate Chocolate Brownies (page 127)
- Coconut Macaroon Bombs (page 124)
- Fermented vegetables
- Gluten-free granola
- Quinoa

MAKE AHEAD

- Basic Vinaigrette (page 135): Store in a glass jar in the refrigerator.
- Homemade Vegan Mayo (page 133): Store in a glass jar in the refrigerator.
- Homemade Almond Milk (page 52): Store in a glass jar in the refrigerator.
- Homemade Ketchup (page 132): Store in a glass jar in the refrigerator.
- Lightly toast pumpkin seeds and almond slivers in a pan: Store in a covered glass jar in the cupboard.
- Fermented Thai Chile Sauce (page 136): Make at least 3 days in advance of starting your diet.
- Pink Cabbage and Garlic Fermented Kimchi (page 62): Make at least 3 days in advance of starting your diet.

Chapter Four

Life After the Gut Health Diet

Congratulations on your seven-day accomplishment! You may have great results to report and wish to continue eating this way as a lifestyle. I certainly recommend it. If you didn't quite experience the improvements that you were hoping for but did notice that you are on the right track, you can take a look at your other options moving forward in this chapter.

Check In with Your Symptoms

Scientists used to think that it would take years to make big gut changes, but a study in *Nature* indicates that these changes can happen incredibly fast in the human gut—often within three or four days of implementing dietary changes. Lawrence David, assistant professor at the Duke Institute for Genome Sciences & Policy, says, "We found that the bacteria that lives in people's guts is surprisingly responsive to change in diet. Within days we saw not just a variation in the abundance of different kinds of bacteria, but in the kinds of genes they were expressing." This is great news for all of us and validation that a seven-day diet can be an impactful start for improved gut health. In seven days, many people on this diet start to notice:

- Normalizing of bowel movements, becoming more productive and perhaps less odorous
- A reduction in sugar and sweet cravings
- More energy at the times of the day when they usually feel a lull or crash
- Lessening of bloating, burping, and other signs of digestive distress; with some symptoms stopping altogether

If you are someone who has had very little to no progress or change in symptoms, this could be indicative of a more complex digestive imbalance, and I would encourage you to repeat the diet in seven-day increments until you begin to see more significant changes. You could also consider taking things up a notch and trying one of the following more restrictive special gut diets to address more complex gut concerns.

Special Gut Health Diets

There are many gut health diets these days, and all have great value, but I chose to highlight the following three diets based on my own experience and because each one claims to help the body heal from markedly different imbalances. These three diets are varied and range in their levels of difficulty; however, they are similar in the following areas:

- Elimination of processed foods
- Elimination of gluten
- Elimination of sugar
- Striving to reduce inflammation in the body

(GAPS) THE GUT AND PSYCHOLOGY SYNDROME DIET

GAPS is a therapeutic diet commonly used in the treatment of inflammatory bowel disease, leaky gut syndrome, autism, ADHD, depression, anxiety, and auto-immune disease.

- There are six phases of the GAPS intro diet, and it takes an average of four to six weeks to complete the phases depending on the progress of the individual.
- The first phase of the intro diet includes broths, soups, boiled meats, fermented foods, and herbal teas.
- The intro diet begins with total avoidance of all wheat and wheat products, packaged foods, fruit juice, dairy, beans, soy, starchy vegetables, and all sugar in every form, including molasses and maple syrup, to name a few.

(BED) THE BODY ECOLOGY DIET

This diet is great for people who are dealing with food intolerances, candida overgrowth, and general digestive discomfort, and for children with autism and behavioral disorders.

- There are two stages of this diet, the first being the most challenging.
- The length of this diet depends on your symptoms as you progress.
- The foods to avoid are all sugar, including all high-glycemic fruit, except sour fruit like lemons and limes, and most grains, except buckwheat, millet, quinoa, and amaranth.
- BED has seven principles to follow:

  1. Balance in contracting and expanding foods

  2. Acid/alkaline balance and body/food choices

  3. Principle of uniqueness when dealing with symptoms and healing progress

  4. Cleansing

  5. Food combining

  6. The 80/20 principle (80 percent alkaline foods/20 percent acidic)

  7. Step-by-step monitoring and journaling every step to track progress and setbacks

LOW-FODMAP DIET

This is a two-phase intervention, with strict reduction of all slowly absorbed or indigestible short-chain carbohydrates, also known as FODMAPs. This is followed by reintroduction of specific FODMAPs according to tolerance. FODMAPs are not the cause of these disorders, but restricting these foods might help to improve digestive symptoms in adults with irritable bowel syndrome (IBS) and other gastrointestinal disorders.

- Dieters need to become very familiar with the Monash University Low FODMAP app and begin a strict elimination diet for at least six to eight weeks; however, starting with this Gut Health Diet for Beginners would likely shorten that process to about four weeks.
- The second phase is the reintroduction process, which consists of adding FODMAP foods back into the diet one group at a time, beginning with a small amount and then gradually increasing the amount of that food group.

How to Choose a Gut Health Diet

I believe these diets all have substantial benefits and can help most people with their health concerns; they all have similar principles and basics. This Gut Health Diet for Beginners is a great way to set you up to transition into any of them, as well as some diets not listed here.

Of the three mentioned, the Low FODMAP diet is the most restrictive, and for that reason, most helpful for people with severe imbalance. I do recommend clinical guidance with a practitioner who is well versed in this diet and its challenges. Its restrictions can leave a person not getting enough nutrients, and as mentioned earlier, the dramatic shift in sugar and processed food intake can create strong "die-off" reactions in the body. It is helpful to have some guidance in such a situation, when you are up against potentially feeling worse before feeling better.

The Body Ecology Diet and the GAPS diet are similar to one another, and they are both geared toward healing the gut and the belief that gut dysbiosis or damaged gut ecology is at the root of most imbalance and disease. They both claim to help bring balance to many inflammatory and autoimmune disorders, as well as such issues as:

- ADD and ADHD
- Autism
- Bipolar disorder
- Childhood bed-wetting
- Depression
- Dyslexia
- Dyspraxia
- Eating disorders
- Gout
- Obsessive-compulsive disorder (OCD)
- Schizophrenia
- Tourette's syndrome

Although clinical guidance is always suggested, you can learn more about these two diets by familiarizing yourself with the books written about them: *The Body Ecology Diet*, by Donna Gates and Linda Schatz, and *Gut and Psychology Syndrome*, by Natasha Campbell-McBride.

They both offer a tremendous amount of guidance as well as resources such as online groups, websites, and additional information.

PART TWO

Recipes for a Healthy Gut

The recipes in this book are all geared toward high-quality, whole-some, homemade, fresh ingredients, organic and non-GMO whenever possible. Almost all of the recipes have no more than 5 ingredients (excluding water, oils, salt, pepper, and other basic pantry items like vinegar, garlic, lemons, limes, etc.) for your cooking and prepping convenience and to lighten the burden on digestion. Where applicable, you'll see one or more of these dietary and recipe tip labels:

VEG Vegetarian or **VEGAN** Vegan

30 MIN The meal can be prepped and cooked in less than 30 minutes.

QUICK PREP The meal does not require more than 10 minutes of preparation.

There will also be tips on how you can modify the recipes for certain restrictions you may have, such as substitutions for Big 8 allergens (wheat, peanuts, tree nuts, fish, shellfish, dairy, soy, and eggs) as well as substitutions for high-FODMAP ingredients and nightshades.

Papaya-Banana Smoothie with Kefir, page 57

Chapter Five

Breakfasts, Smoothies, and Drinks

Cucumber-Avocado Toast with Aleppo Pepper

Serves 1

PREP TIME: 5 MINUTES / COOK TIME: 2 MINUTES

Who doesn't love a fresh, crisp avocado toast? The lime and the Aleppo pepper transform this simple breakfast or snack into a special treat with a savory, mouthwatering burst of flavor. So simple; so delicious.

1 Persian cucumber, thinly sliced lengthwise

Sea salt

Freshly ground black pepper

¼ teaspoon Aleppo pepper (or paprika, crushed red pepper, or chipotle pepper)

1 ripe avocado

½ lime

2 slices Ezekiel bread (see recipe tip)

PER SERVING Calories: 504; Total Fat: 28g; Saturated Fat: 4g; Sodium: 220mg; Carbohydrates: 60g; Fiber: 20g; Protein: 14g

1. Place the cucumber in a medium bowl, and season with salt, black pepper, and the Aleppo pepper.

2. In a small bowl, smash the ripe avocado and season with salt, black pepper, and a squeeze of lime juice.

3. Toast the bread, then spread the avocado onto the toasted bread and top with the cucumber slices. Serve immediately.

STORAGE NOTE: This dish will not keep; however, you can slice and season the cucumber and refrigerate overnight before use. A regular, peeled cucumber can also be used in this recipe. The avocado will brown quickly when exposed to air, so either use immediately or cover and refrigerate the prepared avocado with plastic wrap touching the avocado.

RECIPE TIP: Ezekiel bread is not gluten-free; however, it is the highest quality bread that I will enjoy on occasion (sesame is my favorite kind). You can also make this toast with any gluten-free bread.

Soft Scramble with Pink Cabbage Kimchi

Serves 2

PREP TIME: 2 MINUTES / COOK TIME: 3 MINUTES

I used to think scrambled eggs were so boring until I learned how to make a proper French scrambled egg in culinary school. I've since changed my tune. Buttery, fluffy, delicate goodness. When made this way, they are delightful all on their own, but it's even better to dress them up. Experiment with herbs, lightly sautéed vegetables, and different spices.

½ tablespoon cultured butter or ghee or high-quality butter like Kerrygold

4 large free-range eggs

Sea salt

2 tablespoons Pink Cabbage and Garlic Fermented Kimchi (page 62)

PER SERVING Calories: 170; Total Fat: 13g; Saturated Fat: 5g; Sodium: 215mg; Carbohydrates: 1g; Fiber: 0g; Protein: 13g

1. In a medium skillet over medium-low heat, melt the butter to coat the whole pan.

2. In a medium bowl, whisk the eggs until bubbles form. When the butter is bubbling lightly, add the eggs to the center of the pan.

3. Using a spatula, gently fold the eggs over each other until curds start to form. Don't allow the eggs to set on the edges; continue to fold them quickly. Do not overcook; they should be large, moist folds, not separated at all.

4. Slide the eggs onto a plate, and sprinkle with sea salt.

Continued

Soft Scramble with Pink Cabbage Kimchi *Continued*

5. Garnish with the Pink Cabbage and Garlic Fermented Kimchi and serve.

STORAGE NOTE: These eggs are best eaten immediately; however, you can refrigerate for 2 days for a nice egg salad later.

RECIPE TIP: Eggs are one of the most complete, highly nutritious foods available, but they are also one of the top 8 allergens. I encourage anyone who suspects a sensitivity to eggs to go without them for 10 days and then reintroduce them, checking for any reaction. If you suspect any reaction to eggs, it is best to opt for other breakfast options in this book.

Rainbow Vegetable Hash and Soft-boiled Egg

Serves 4

PREP TIME: 15 MINUTES / COOK TIME: 15 MINUTES

This is a versatile dish that can be served alone or as a side for breakfast, lunch, or dinner. These starchy veggies are also a great snack when you're feeling that midday drag. I like to slice half an avocado on top, and it's instant energy. I have also found that a couple of tablespoons of starchy veggies before bed helps me sleep like a baby.

4 free-range eggs

2 large Yukon Gold or russet potatoes, peeled and chopped into small, bite-size cubes

1 small sweet potato or garnet yam, peeled and chopped into small, bite-size cubes

2 tablespoons coconut oil or avocado oil

1 large red bell pepper, cored, seeded, and diced

1 medium onion, diced small

2 garlic cloves, minced or ¼ teaspoon granulated garlic or garlic powder

Sea salt

Freshly ground black pepper

PER SERVING Calories: 246; Total Fat: 11g; Saturated Fat: 7g; Sodium: 96mg; Carbohydrates: 29g; Fiber: 5g; Protein: 9g

1. In a small saucepan, cover the eggs with water. Over medium-high heat, bring the water to a boil and set a timer for 3 minutes. When the time is up, drain, immediately run the eggs under cool water, and set aside.

2. In the saucepan, combine the potato and sweet potato cubes and cover with water. Bring to a boil, then reduce the heat and simmer until you can pierce the potatoes easily with a fork but they don't fall apart, 5 to 7 minutes. Drain the potatoes and dry with a paper towel. Set aside.

Continued

Rainbow Vegetable Hash and Soft-boiled Egg *Continued*

3. Meanwhile, in a large skillet, heat the oil over medium-high heat. Add the bell pepper, onion, garlic, and precooked potatoes to the pan and cook, occasionally turning the veggies, until cooked through but still firm and crispy, 3 to 5 minutes. Sprinkle with salt and pepper to taste. Keep the hash warm.

4. Peel the eggs and place one egg on each serving of hash, slice the egg with a butter knife, and serve.

STORAGE NOTE: The hash can be refrigerated for 1 week.

RECIPE TIP: If you are vegan or have any egg allergies or sensitivities to eggs, simply omit the egg. I also love to add fresh arugula to the warm hash, as it will lightly wilt the greens and add a peppery crunch. Of course, a side of Pink Cabbage and Garlic Fermented Kimchi (page 62) is also a welcome flavor to add in.

Country Frittata with Feta Cheese

Serves 4

PREP TIME: 10 MINUTES / COOK TIME: 25 MINUTES

This is a fantastic dish to serve for an old-fashioned Sunday brunch. The feta cheese adds a sharp tang to the earthiness of this comfort dish. Truly one of my favorites, and maybe now one of yours, too.

2 Yukon Gold potatoes, peeled and chopped into bite-size cubes

Sea salt

2 tablespoons coconut oil, avocado oil, or high-quality butter

2 large handfuls spinach

1 red bell pepper, cored, seeded, and diced

12 free-range eggs

½ cup crumbled feta cheese

PER SERVING Calories: 383; Total Fat: 24g; Saturated Fat: 13g; Sodium: 472mg; Carbohydrates: 21g; Fiber: 3g; Protein: 22g

1. Preheat the oven to 350°F.

2. Place the potatoes in a small sauce-pan with a sprinkle of salt, and cover with water. Bring to a boil, then reduce the heat and simmer until the potatoes pierce easily with a fork but they don't fall apart, 5 to 7 minutes. Drain the potatoes and dry with a paper towel. Set aside.

3. In a large, oven-safe skillet over medium heat, heat the oil. Add the spin-ach and cook down until the leaves are still formed and not totally wilted, about 3 minutes.

4. Add the bell pepper and cook until the pepper is tender but still firm, about 5 minutes.

5. Crack all the eggs into a large bowl, and whisk until bubbles form. Add the eggs into the center of the pan, and mix until everything is coated with the egg, then turn off the heat.

Continued

Country Frittata with Feta Cheese *Continued*

6. Sprinkle the mixture with the feta cheese.

7. Place the skillet on the center of the middle oven rack and cook for 20 minutes, or until the top is browning. Slice and serve.

STORAGE NOTE: This frittata keeps well for lunches and snacks and can be refrigerated for 1 week.

RECIPE TIP: You can really get creative with this frittata and add all sorts of yummy vegetables—use whatever you've got on hand.

Spaghetti Squash Hash Browns and Soft-boiled Egg

Serves 2

PREP TIME: 10 MINUTES / COOK TIME: 45 MINUTES

These crispy squash hash browns are so satisfying, especially if you're craving French fries. Crispy, salty, and starchy—they meet all these criteria. Squash is also a good alternative to pasta when a craving arises; it's a great source of fiber, which is basically food for good bacteria in the gut. A high–fiber diet is very important for a healthy gut.

1 medium spaghetti squash

2 tablespoons avocado oil or coconut oil, plus more for oiling the squash

Sea salt

1 free-range egg

Freshly ground black pepper

1 soft-boiled free-range egg

PER SERVING Calories: 267; Total Fat: 20g; Saturated Fat: 13g; Sodium: 230mg; Carbohydrates: 18g; Fiber: 0g; Protein: 8g

1. Preheat the oven to 400°F.

2. Halve the spaghetti squash lengthwise, and scoop out and discard the seeds.

3. Drizzle a little avocado oil and sprinkle a little salt on the inside of the squash.

4. Place the squash, flat-side down, on a sheet pan. Roast for 40 minutes, or until you can slide a knife into the squash easily.

5. Using a fork, fluff the inside of the squash into a spaghetti-like consistency. Scoop the strands into a large bowl and set aside until cool.

Continued

Spaghetti Squash Hash Browns and Soft-boiled Egg *Continued*

6. Once cool, place the squash in a paper towel or use clean hands over the sink, and squeeze all the moisture out of the squash.

7. Return the squash to the bowl, and add one egg and salt and pepper to taste.

8. Using clean hands, mix the ingredients thoroughly. Form into little patties.

9. In a large skillet over medium heat, heat the oil until shimmering.

10. Fry the squash patties, flipping once, until golden brown on both sides, about 5 minutes total.

11. Serve the hash browns with a soft-boiled egg on top.

STORAGE NOTE: The hash browns will keep, refrigerated, for 1 to 2 weeks.

RECIPE TIP: This is one of those dishes that can be premade and makes life easier. Either cook the squash ahead of time and refrigerate the "spaghetti" until you are ready to make the hash browns or premake the hash browns, and just heat them back up in a pan and serve.

Brown Rice Porridge

Serves 1

PREP TIME: 5 MINUTES / COOK TIME: 20 MINUTES

A comforting, filling way to start your day, or enjoy as a snack when you get home from work. Brown rice is one of the easiest grains on the gut. You can get creative with this porridge by adding slivered almonds, shaved coconut, golden berries, blueberries, walnuts—the possibilities are endless.

2 cups water

1 cup organic short-grain brown rice

1 (13.5-ounce) can unsweetened coconut milk

2 tablespoons pure maple syrup

2 teaspoons ground cinnamon

PER SERVING Calories: 1,166; Total Fat: 93g; Saturated Fat: 81g; Sodium: 61mg; Carbohydrates: 90g; Fiber: 14g; Protein: 12g

1. In a medium saucepan, bring the water to a high boil and add the rice.

2. Lower the heat to simmer.

3. Add the coconut milk, syrup, and cinnamon and cover, stirring occasionally.

4. Turn off the heat when the liquid has been absorbed and the rice is tender, about 20 minutes, depending on desired doneness, and serve.

STORAGE NOTE: This dish is best eaten immediately.

RECIPE TIP: This can easily be prepared in a slow cooker. Simply add the rice, water, coconut milk, cinnamon, and syrup. Cook on low for 3½ hours, or until the liquid is absorbed. Also, soaking the rice overnight before cooking it can help with digestion, and if you'd like to modify the dish to contain no sugar, you can replace the maple syrup with stevia or yacon syrup.

Homemade Almond Milk

Makes 4 cups

PREP TIME: 10 MINUTES

I use almond milk for so many smoothie recipes, and you will soon discover that store-bought almond milk is a shadow of a wholesome homemade version. There also can be many hidden ingredients in most almond milks, so if you choose to buy them, inspect the labels carefully.

2 cups raw almonds (soaked in water for at least 20 minutes)

4 cups water

PER SERVING (1 cup) Calories: 30; Total Fat: 3g; Saturated Fat: 0g; Sodium: 22mg; Carbohydrates: 2g; Fiber: 1g; Protein: 1g

1. In a blender, liquefy the almonds and water.

2. Strain the mixture through a fine mesh strainer, nut milk bag, or cheesecloth into a bowl or quart jar, and discard the almond pulp.

3. Store in a tightly sealed container in the refrigerator.

STORAGE NOTE: Homemade almond milk will keep refrigerated for 2 weeks.

RECIPE TIP: Nut milk bags can be purchased online or in specialty stores. Anyone with a tree nut allergy should avoid almonds and opt for dairy kefir, or simply substitute water in smoothies that call for nut milks.

Golden Turmeric Milk

Serves 2

PREP TIME: 5 MINUTES / COOK TIME: 10 MINUTES

This luscious milk is filled with healthy properties for the gut and the rest of the body. It's calming and soothing for the lining of the stomach and intestinal wall and brings down inflammation throughout the gut. A great way to start the day as a coffee substitute, or enjoy after a meal before bed.

2 cups unsweetened coconut milk

1-inch piece fresh turmeric, peeled and sliced thinly, or 1 teaspoon ground turmeric

½-inch piece fresh ginger, peeled and sliced, or ½ teaspoon ground ginger

2 tablespoons coconut oil

½ teaspoon raw honey

1 cinnamon stick

Dash freshly ground black pepper

PER SERVING Calories: 173; Total Fat: 18g; Saturated Fat: 16g; Sodium: 1mg; Carbohydrates: 5g; Fiber: 1g; Protein: 0g

1. In a small saucepan over medium-low heat, bring the coconut milk, turmeric, ginger, coconut oil, honey, cinnamon stick, and pepper to a low simmer for 10 minutes.

2. Strain through a mesh strainer into a blender and blend until frothy.

STORAGE NOTE: Refrigerate in a sealed container for up to 2 weeks.

RECIPE TIP: This recipe can be upgraded using other exotic spices like cardamom, cloves, and nutmeg, and even a teaspoon of ghee or high-quality butter.

Alkalizing Green Smoothie

Makes 2 cups

PREP TIME: 5 MINUTES

This alkalizing smoothie is chock-full of minerals and vitamins that help support immunity, alkalize your system, and reduce inflammation, ultimately supporting gut health. Add some high-quality greens that have probiotics for even more gut support. This should be a staple in your household—I recommend enjoying it weekly, if not daily. There is nothing more important to gut health than concentrated absorbable minerals in your diet. And since it's called a smoothie, you want it to be very smooth with a drinkable consistency.

1 handful romaine lettuce

1 handful spinach

½ ripe banana (frozen optional for creamier consistency)

½ small lemon with skin, washed thoroughly

1 cup ice

PER SERVING (2 cups) Calories: 82; Total Fat: 0g; Saturated Fat: 0g; Sodium: 52mg; Carbohydrates: 20g; Fiber: 4g; Protein: 3g

In a high-powered blender, blend to combine the romaine, spinach, banana, lemon, and ice until smooth.

STORAGE NOTE: This drink can be refrigerated in a sealed container for 24 hours, but it's best consumed immediately. Consider buying a mini blender for the office; green smoothies on the go can make a world of difference.

RECIPE TIP: Experiment with frozen raspberries, blueberries, or other low-glycemic fruits. It's also a good idea to rotate your greens—one day romaine, the next kale, the next spinach—to diversify the benefits. Including healthy additions like flaxseed and essential fats will make this a super shake.

Berry 'n' Greens Smoothie

Makes 2 cups

PREP TIME: 5 MINUTES

Not a fan of green vegetables? There's no better way to hide them than in this deliciously deceiving berry smoothie.

½ cup frozen blueberries

½ cup frozen raspberries

1 handful spinach

½ banana

½ cup water or Homemade Almond Milk (page 52)

PER SERVING (2 cups) Calories: 138; Total Fat: 1g; Saturated Fat: 0g; Sodium: 49mg; Carbohydrates: 35g; Fiber: 8g; Protein: 3g

In a high-powered blender, blend to combine the blueberries, raspberries, spinach, banana, and water until smooth. Add more fluid if needed, and blend again.

STORAGE NOTE: This drink can be refrigerated in a sealed container for 24 hours. But this smoothie is best consumed immediately.

RECIPE TIP: Sticking with low-glycemic fruit like berries is a great option when trying to avoid sugar.

Monkey Madness Smoothie

Makes 2 cups

PREP TIME: 5 MINUTES

I am crazy about this filling, sweet, nutritious smoothie. It's perfect for those days when you are craving something sweet, and it's great as a snack, dessert, and even breakfast.

1 cup unsweetened almond milk (preferably homemade, page 52)

1 frozen banana

½ cup coconut water

2 or 3 dried Medjool dates, pitted

2 tablespoons almond butter

PER SERVING **(2 cups)** Calories: 419; Total Fat: 23g; Saturated Fat: 2g; Sodium: 45mg; Carbohydrates: 55g; Fiber: 7g; Protein: 8g

In a high-powered blender, blend to combine the almond milk, banana, coconut water, dates, and almond butter until smooth.

STORAGE NOTE: This drink can be refrigerated in a sealed container for 24 hours. But this smoothie is best consumed immediately.

RECIPE TIP: If you have a tree nut allergy, you can replace the nut butter and almond milk with water.

Papaya-Banana Smoothie with Kefir

Makes 4 cups

PREP TIME: 5 MINUTES

This smoothie is filled with ingredients that are diversely beneficial for gut health. Papaya has many gut benefits; in fact, people in some tropical countries use it as a treatment for IBS and other bowel disorders. It's also abundant in antioxidants, vitamin C, and one of the gut's best friends, fiber.

1 cup fresh or frozen papaya (if using fresh papaya, set seeds aside)

1 cup fresh or frozen pineapple

1 banana

1 cup unsweetened almond milk (preferably homemade, page 52)

¾ cup plain goat or cow kefir (preferably homemade, page 58)

PER SERVING (2 cups) Calories: 186; Total Fat: 3g; Saturated Fat: 1g; Sodium: 144mg; Carbohydrates: 38g; Fiber: 6g; Protein: 6g

In a high-powered blender, blend to combine the papaya, pineapple, banana, almond milk, and kefir until smooth.

STORAGE NOTE: This drink can be refrigerated in a sealed container for 24 hours. But this smoothie is best consumed immediately.

RECIPE TIP: To add even more gut-boosting benefit, try adding half a thumb of peeled fresh ginger, a squeeze of lime, and 1 tablespoon of papaya seeds. If you have sensitivities to dairy, simply omit the kefir, or unsweetened coconut yogurt is a great substitute.

Homemade Kefir

Makes 4 cups

You can't have a diet specific to gut health without fermented foods! Kefir is one of the most powerful and yummy fermented foods and has been eaten for centuries. It's kind of like a watery yogurt: sour, tangy, and a bit fizzy. You can acquire kefir grains if you choose, but these days you can find kefir starters, products like Yogourmet—my favorite is the Body Ecology kefir starter, which can be purchased online. These bacteria are designed to specifically aid in gut health. As a general rule, ½ cup of kefir before bed is recommended.

4 cups organic, low-fat, unpasteurized milk

½ cup organic heavy (whipping) cream

1 packet kefir starter or yogurt starter

PER SERVING (½ cup) Calories: 103; Total Fat: 7g; Saturated Fat: 4g; Sodium: 62mg; Carbohydrates: 7g; Fiber: 0g; Protein: 4g

1. In a medium saucepan, heat the milk and cream to 92°F (31° or 32°C). Be careful not to overheat. If you don't have a thermometer, use your finger to check the temperature. It should feel neutral to the touch, not hot or cold.

2. Pour the warmed milk mixture into a blender.

3. Empty the kefir starter packet into the blender.

4. Blend a few seconds to mix the milk and the starter.

5. Pour the liquid into a glass jar. Top with a tight-fitting lid.

6. Store at room temperature to ferment, ideally 72°F to 75°F for 18 to 24 hours. It will thicken until slightly clumpy and have a distinctly sour aroma.

7. Refrigerate.

STORAGE NOTE: This drink can be refrigerated in a sealed container for up to 2 weeks.

RECIPE TIP: If you find the taste of kefir slightly gamey, you can mask it with cinnamon and blend with fresh fruits. Instead of using a new kefir starter packet every time you make kefir, simply use ¼ cup of this first batch to reinoculate future batches (up to 7 times). This makes purchasing these starters very economical compared to purchasing a high-quality, over-the-counter probiotic supplement, and according to recent research, far more effective. You can use organic coconut water instead of dairy if you are vegan or have any sensitivities to dairy.

Roasted Balsamic Green Veggies and Mushrooms, page 66

Chapter Six

Vegetables and Salads

Pink Cabbage and Garlic Fermented Kimchi

— Makes about 8 cups or 2 (1-quart) jars —

PREP TIME: 30 MINUTES, PLUS 3 TO 10 DAYS TO FERMENT

This is my take on the famous and pungent Korean kimchi. This crunchy, bright version of a classic dish goes with almost everything: eggs, veggies, meat, guacamole, and there's nothing wrong with sitting down to a bowl all by itself.

3½ cups filtered (not tap) water

1 teaspoon chili flakes (cayenne, crushed red pepper flakes, or Korean peppers)

6 garlic cloves, 3 minced, 3 whole peeled, divided

1 tablespoon sea salt

1 packet culture starter (optional)

2 medium cabbages, one green, one purple

PER SERVING (1 cup) Calories: 60; Total Fat: 0g; Saturated Fat: 0g; Sodium: 536mg; Carbohydrates: 14g; Fiber: 6g; Protein: 3g

1. In a blender, combine the water, chili flakes, 3 whole garlic cloves, salt, and the culture starter (if using). Liquefy and set aside.

2. Remove the cores from the cabbages and shred with a food processor or use a chef's knife to cut into ribbon-like pieces.

3. In a large bowl, mix together the blender contents with the cabbage and minced garlic, massaging with your hands until tender and juicy.

4. Pack the cabbage tightly in glass jars, adding all the excess juice into the jars.

5. Store at 72°F for at least 3 days and preferably 7 days, for a crunchier, spicier flavor.

STORAGE NOTE: If the jar remains closed, the ferment will last at room temperature for years! Once opened, refrigerate immediately and use within a month. The jars tend to leak as they ferment, so setting them in a large plastic container or on a rimmed baking sheet is ideal. They also give off a pungent odor, which is entirely normal. Keeping them warm is the most important factor for a good ferment (at least 72°F but no warmer than 84°F).

RECIPE TIP: You should adjust your spice and pungency levels according to your liking. I am a fan of very pungent flavors and mild to strong spice. The more garlic and chili you add, the stronger it will be. Fermenting vegetables is all about experimenting and can be done with just about any veggie you can think of. I recommend using a heating pad to help keep your ferments from dropping below 72°F. Also, I like to use Body Ecology brand culture starter.

Roasted Maple Brussels Sprouts and Pumpkin Seeds

Makes 4 cups

PREP TIME: 5 MINUTES / COOK TIME: 30 MINUTES

Some people say, "You either love Brussels sprouts or you hate them." I say, if you roast them with maple syrup, everyone loves them.

1 pound Brussels sprouts, ends cut off, halved

2 tablespoons pure maple syrup

2 tablespoons coconut oil or avocado oil

Sea salt

Freshly ground black pepper

½ cup Toasted Pumpkin Seeds with Sea Salt (page 130)

PER SERVING (1 cup) Calories: 227; Total Fat: 15g; Saturated Fat: 8g; Sodium: 91mg; Carbohydrates: 20g; Fiber: 5g; Protein: 8g

1. Preheat the oven to 400°F.

2. On a rimmed baking sheet, toss together the Brussels sprouts, maple syrup, oil, salt, and pepper.

3. Roast the Brussels sprouts for 30 minutes, stirring occasionally, until still firm but pierced easily with a knife.

4. Serve sprinkled with the Toasted Pumpkin Seeds.

STORAGE NOTE: This dish will keep, refrigerated, for 1 week.

RECIPE TIP: If Brussels sprouts are undercooked, they can be harder to digest and cause bloating or gas, so make sure they are easily pierced with a knife or fork when roasting.

Roasted Cauliflower with Slivered Almonds

Serves 4

PREP TIME: 5 MINUTES / COOK TIME: 20 MINUTES

This dish is simple, but juicy and crunchy. It's a wonderful side to accompany any protein and can be easily transformed into a soup by adding a few cups of vegetable broth and liquefying in a blender.

1 large cauliflower head

2 tablespoons coconut oil or avocado oil

Sea salt

Freshly ground black pepper

½ cup toasted almond slivers (see page 31)

PER SERVING Calories: 132; Total Fat: 9g; Saturated Fat: 6g; Sodium: 123mg; Carbohydrates: 12g; Fiber: 6g; Protein: 5g

1. Preheat the oven to 400°F.

2. Chop the cauliflower into small florets.

3. On a large, rimmed baking sheet, toss the cauliflower with the oil, salt, and pepper.

4. Roast for 20 minutes, or until the florets are easily pierced with a knife.

5. Serve sprinkled with the toasted almonds.

STORAGE NOTE: This dish will keep, refrigerated, for 1 week.

RECIPE TIP: Cauliflower can be quite bland unless you use plenty of seasonings and cook it through well. Don't hesitate to add plenty of sea salt and other spices. For anyone with tree nut allergies, simply omit the almonds and instead sprinkle with Toasted Pumpkin Seeds with Sea Salt (page 130).

Roasted Balsamic Green Veggies and Mushrooms

Serves 4

PREP TIME: 20 MINUTES / COOK TIME: 20 MINUTES

Roasted vegetables are the simplest, most yummy dish. If you don't undercook them, they are easily digested, and if you've never been a big vegetable lover, you can also smother them with spices or even my Healthy BBQ Sauce (page 137). Grandma was spot on when she said, "Eat your veggies"—there are more minerals and vitamins in this dish than in any supplement you could possibly buy.

½ pound green beans, trimmed and cut in half

4 broccoli florets

5 mushrooms, cut into bite-size pieces (cremini or shiitake are great choices)

4 Brussels sprouts, cut into bite-size pieces

¼ cup balsamic vinegar

1 teaspoon sea salt

¼ teaspoon freshly ground black pepper

3 tablespoons coconut oil, avocado oil, or raw sesame oil

PER SERVING Calories: 130; Total Fat: 11g; Saturated Fat: 9g; Sodium: 486mg; Carbohydrates: 8g; Fiber: 4g; Protein: 3g

1. Preheat the oven to 420°F.

2. On a large, rimmed baking sheet, toss the green beans, broccoli, mushrooms, Brussels sprouts, vinegar, salt, pepper, and coconut oil until well coated.

3. Roast for 20 minutes, or until tender and browning on the edges, and serve.

STORAGE NOTE: This dish will keep, refrigerated, for 3 days.

RECIPE TIP: This is one of those dishes that I make in advance, refrigerate, and have handy for any meal. I even found a container of this that had been hiding for 3 or 4 days in the back of the fridge, heated them up, blended them with some vegetable broth, and, voilà—I had vegetable soup!

Roasted Root Vegetable Soup

Makes 4 cups

PREP TIME: 15 MINUTES / COOK TIME: 30 MINUTES

This soup is a comforting, easy-to-digest dish that can serve as a main or a side. You can also pour it over quinoa or wild rice. And soup, in my opinion, is the most overlooked breakfast option.

1 large garnet yam or sweet potato, peeled and roughly chopped

2 carrots, roughly chopped

1 large parsnip, roughly chopped

1 onion, roughly chopped

¼ cup coconut oil or avocado oil

1 teaspoon sea salt

¼ teaspoon freshly ground black pepper

1 cup unsweetened coconut milk, at room temperature

4 cups water or less, as needed for desired consistency

PER SERVING (1 cup) Calories: 345; Total Fat: 28g; Saturated Fat: 25g; Sodium: 519mg; Carbohydrates: 24g; Fiber: 6g; Protein: 4g

1. Preheat the oven to 400°F.

2. On a large, rimmed baking sheet, toss the yam, carrots, parsnip, and onion with the oil.

3. Roast the vegetables for 20 to 30 minutes, or until easily pierced with a knife.

4. In a blender, combine the roasted vegetables with the salt, pepper, and coconut milk and liquefy. Add water as needed, blend again, and serve.

STORAGE NOTE: This soup will keep, refrigerated, for 1 week, or frozen for longer.

RECIPE TIP: This is a great dish to make in large quantities and freeze in individual containers. Thaw one container in the sink overnight. Warm the soup in the morning, and off you go to work.

Arugula and Warm Squash Salad with Pomegranate Seeds and Feta

——— *Serves 3* ———
PREP TIME: 10 MINUTES / COOK TIME: 20 MINUTES

This fall salad is light and tangy, and the pomegranate seeds add crunch. Once again, here's a dish that goes with any main course or serves as a salad on its own for lunch.

1 (2- to 3-pound) butternut squash, peeled and chopped into bite-size cubes

¾ cup coconut oil or avocado oil

Sea salt

Freshly ground black pepper

3 handfuls arugula

½ cup loose pomegranate seeds

½ cup feta cheese, crumbled

2 tablespoons freshly squeezed lemon juice

1 (3- to 4 ounce) grilled chicken breast (optional)

—————

PER SERVING Calories: 501; Total Fat: 42g; Saturated Fat: 35g; Sodium: 372mg; Carbohydrates: 29g; Fiber: 5g; Protein: 6g

1. Preheat the oven to 420°F.

2. On a large, rimmed baking sheet, toss the butternut squash with the oil, salt, and pepper, and spread evenly.

3. Roast the squash for 20 minutes, or until tender and beginning to brown.

4. Meanwhile, in a large bowl, toss the arugula, pomegranate seeds, feta, and lemon juice.

5. Once the squash is roasted, carefully pour the squash over the salad, give it another toss, and serve.

6. If using, slice the grilled chicken breast into long strips and serve atop the salad, cold or warm.

STORAGE NOTE: Because the squash is served warm over the salad, it will wilt the greens, so it's best served and eaten immediately.

RECIPE TIP: Vegans and those who are dairy-free can easily omit the feta from this recipe and add toasted almond slivers (page 31), Toasted Pumpkin Seeds with Sea Salt (page 130), and garlic as great additions for more flavor.

Quinoa-Spinach Salad with Toasted Almond Slivers

Serves 2 to 3

PREP TIME: 5 MINUTES / COOK TIME: 10 MINUTES

Quinoa is one of the best grains for gut health, because it is not a grain at all, but rather a grain-like seed. This just means it doesn't contain the same properties as in regular grains that make them hard to digest. This fluffy, crunchy, lemony salad goes great as a side dish to just about any animal protein, and it's great alone as a light lunch or a quick healthy snack. For a more filling meal, chop up some chicken or top with a piece of grilled salmon.

2 cups salted water

½ cup rinsed quinoa

2 cups chopped spinach

½ cup diced cucumbers

1 tablespoon freshly squeezed lemon juice

1½ tablespoons coconut oil or avocado oil

Sea salt

Freshly ground black pepper

¼ cup toasted almond slivers (see page 31)

PER SERVING (half the recipe) Calories: 326;
Total Fat: 19g; Saturated Fat: 10g; Sodium: 145mg;
Carbohydrates: 32g; Fiber: 5g; Protein: 10g

1. In a medium saucepan, boil the salted water. Add the quinoa and boil for about 10 minutes, until tender. Drain and let cool.

2. In a large bowl, mix together the spinach, cucumbers, and quinoa.

3. To make the dressing, in a small bowl, whisk the lemon juice, oil, salt and pepper. Pour the dressing onto the salad and toss.

4. Sprinkle with the toasted almond slivers and serve.

STORAGE NOTE: This salad will keep, refrigerated, for 1 week.

RECIPE TIP: For anyone with tree nut allergies, simply omit the almond slivers and substitute some Toasted Pumpkin Seeds with Sea Salt (page 130).

Greek Quinoa Tabbouleh

Serves 4

PREP TIME: 10 MINUTES / COOK TIME: 10 MINUTES

This summery, light salad is so easy. Turn up the flavor with some feta cheese, toasted almonds, or scallions. It's a versatile dish that keeps well for lunch or picnics.

2 cups salted water

½ cup rinsed quinoa

2 finely minced garlic cloves

2 tablespoons coconut oil or avocado oil

2 tablespoons freshly squeezed lemon juice

Sea salt

Freshly ground black pepper

2 cups diced cucumbers

3 tablespoons minced fresh parsley

PER SERVING Calories: 150; Total Fat: 8g;
Saturated Fat: 6g; Sodium: 64mg; Carbohydrates: 16g;
Fiber: 2g; Protein: 4g

1. In a medium saucepan, boil the water. Add the quinoa and boil for about 10 minutes, until tender. Drain and let cool.

2. In a small bowl, mix the garlic, oil, lemon juice, salt, and pepper.

3. In a large bowl, combine the cucumbers, parsley, and quinoa. Toss together with the dressing and serve.

STORAGE NOTE: This salad will keep, refrigerated, for 1 week.

RECIPE TIP: These quinoa salads can be premade and used as sides, mains, or snacks.

Wild Rice Stir-Fry

Serves 4

PREP TIME: 10 MINUTES / COOK TIME: 15 MINUTES

Wild rice is hearty and surprisingly easy to digest. Adding a couple of free-range eggs when sautéing the vegetables adds a whole new and delicious level to this dish.

2 cups wild rice

1 tablespoon coconut oil

1 cup shredded cabbage

½ cup onion, finely chopped

1 tablespoon crushed garlic

1 teaspoon coconut aminos or gluten-free soy sauce (also called tamari)

1 teaspoon toasted sesame oil

Sea salt

Freshly ground black pepper

PER SERVING Calories: 339: Total Fat: 5g;
Saturated Fat: 3g; Sodium: 92mg; Carbohydrates: 63g;
Fiber: 6g; Protein: 12g

1. Make the rice per package directions. Set aside.

2. In a large skillet over medium heat, melt the coconut oil.

3. Add the cabbage, onion, garlic, coconut aminos, and sesame oil, and season with salt and pepper. Sauté, stirring, until the cabbage and onion are translucent, about 15 minutes.

4. In a large bowl, toss together the rice and cooked vegetables and serve.

STORAGE NOTE: This dish will keep, refrigerated, for 1 week.

RECIPE TIP: I believe the more veggies you can pack into this dish, the better. It's also never a bad idea to soak any grain overnight to ensure any hard-to-digest components are removed.

Baby Red Potato Salad

Serves 4

PREP TIME: 10 MINUTES, PLUS 1 HOUR TO CHILL / COOK TIME: 20 MINUTES

The first thing people think when they are on a "diet" is that the food won't be comforting, filling, or flavorful. Well, this potato salad has a thing or two to teach other diets. It's filled with comfort, it's filling, and it's very flavorful. It can be eaten as a main meal; it's that delicious and nutritionally rich. This combination of ingredients is also incredibly easy on digestion.

2 pounds small red potatoes, washed and cut into bite-size cubes

Sea salt

¾ cup Homemade Vegan Mayo (page 133)

½ cup red onion, finely chopped

¼ cup minced fresh dill

½ cup minced fresh parsley

PER SERVING Calories: 344; Total Fat: 15g; Saturated Fat: 2g; Sodium: 338mg; Carbohydrates: 50g; Fiber: 6g; Protein: 5g

1. In a large saucepan, cover the potatoes with water and cook until tender, about 20 minutes. Salt to taste.

2. Chill the potatoes in the refrigerator for 1 hour.

3. Once chilled, in a large bowl, toss the potatoes together with the Homemade Vegan Mayo, red onion, dill, and parsley until well coated and serve.

STORAGE NOTE: This salad will stay fresh in the refrigerator for 1 week.

RECIPE TIP: These ingredients can be prepared in advance and refrigerated until ready for combining. This is also a great recipe to make in large quantities and bring to a barbecue. If you have a tree nut allergy, just use the recipe tip on page 133 to switch the almond-based vegan mayo to a nut-free version.

Baked Russet French Fries with Parsley and Garlic

Serves 4

PREP TIME: 15 MINUTES / COOK TIME: 35 MINUTES

This is such a go-to in my household. I probably make some sort of French fry three times a week. They are one of the best sources of prebiotic fiber for the gut and easy as can be to make. Spice them up with cayenne or paprika or even chipotle pepper for a different take on the flavor.

2 pounds russet potatoes

2 tablespoons coconut oil or avocado oil

1 teaspoon sea salt

½ teaspoon freshly ground black pepper

¾ cup finely grated Parmesan cheese

½ cup parsley, finely chopped

2 garlic cloves, crushed

PER SERVING Calories: 288; Total Fat: 12g; Saturated Fat: 9g; Sodium: 681mg; Carbohydrates: 38g; Fiber: 8g; Protein: 11g

1. Preheat the oven to 420°F.

2. Peel and cut the potatoes into ½-inch-thick wedges.

3. Place the potatoes on a rimmed baking sheet and toss with the oil, salt, and pepper.

4. Bake for 30 minutes, then sprinkle with the Parmesan, parsley, and garlic.

5. Bake for 5 more minutes, or until the garlic is lightly toasted and the Parmesan is melting. Remove from the oven and serve.

STORAGE NOTE: These French fries will keep, refrigerated, for 24 hours.

RECIPE TIP: Vegans or those with dairy sensitivities can omit the Parmesan cheese. This recipe is great for sweet potato fries; just use garnet yams instead of russets.

Buttery Yams and Walnuts

Serves 3

PREP TIME: 5 MINUTES / COOK TIME: 45 MINUTES

This decadent, silky, sweet dish tastes almost sinful. You can easily double or triple this recipe and serve it for a holiday side, or at a picnic or barbecue. Yams and sweet potatoes are some of the easiest foods to digest, and they are wholesome and great for the gut with their high fiber and vitamin A content.

3 large garnet yams or sweet potatoes, scrubbed

4 tablespoons high-quality butter (I like Kerrygold)

¼ teaspoon sea salt

½ cup chopped walnuts

PER SERVING Calories: 281: Total Fat: 19g;
Saturated Fat: 10g; Sodium: 337mg; Carbohydrates: 27g;
Fiber: 4g; Protein: 3g

1. Preheat the oven to 420°F.

2. On a large, rimmed baking sheet, bake the yams for 45 minutes, or until you can easily pierce each with a kitchen knife.

3. Using a potholder, hold the yam still as you run a knife down the middle, cutting each one in half.

4. Using a large spoon, scoop the yams into a large bowl.

5. Stir in the butter, salt, and walnuts, folding gently until mixed thoroughly, and serve.

STORAGE NOTE: This dish will keep, refrigerated, for 1 week.

RECIPE TIP: If you have a dairy allergy, omit the butter. These yams are absolutely delicious on their own, but this dish is easily upgraded with cinnamon, cardamom, golden berries, or nutmeg—or try them all.

Shrimp Scampi, page 79

Chapter Seven

Seafood and Meatless Mains

Baked Salmon in a Foil Pouch

Serves 2

PREP TIME: 10 MINUTES / COOK TIME: 20 MINUTES

This style of cooking is great for grilling, camping, and impressing friends. It's clean, easy to make, and results in a juicy, tasty dish.

2 tablespoons ghee or coconut oil, divided

2 (6- to 8-ounce) salmon fillets, skin-on

8 asparagus spears, washed and cut into bite-size pieces

1 medium lemon, ½ juiced, ½ cut into thin, round slices

1 tablespoon minced fresh dill (dried is also fine)

Sea salt

Freshly ground black pepper

PER SERVING Calories: 355; Total Fat: 17g; Saturated Fat: 8g; Sodium: 462mg; Carbohydrates: 5g; Fiber: 2g; Protein: 49g

1. Preheat the oven to 400°F.

2. On each of two large pieces of aluminum foil, place 1 tablespoon of ghee.

3. Place one piece of fish on each piece of foil.

4. Distribute the chopped asparagus between the two pieces of foil.

5. Sprinkle the lemon juice and minced dill on top of the vegetables and fish. Add salt and pepper to taste. Top each fillet with the lemon slices.

6. Fold the edges of the foil like an envelope, making sure nothing can leak out.

7. Bake for about 20 minutes, or until the fish is flaky and tender, and serve.

STORAGE NOTE: This dish will keep, refrigerated, for 2 to 3 days.

RECIPE TIP: Cooked pouches are hot, so use oven mitts when handling. If you have any sensitivities to fish, you can use chicken in its place.

Shrimp Scampi

Serves 2

PREP TIME: 10 MINUTES / COOK TIME: 5 MINUTES

I come back to this recipe over and over, because it's clean, bright, no fuss, and easy to digest. A buttery, lemony bite of shrimp pairs with some red pepper for a little kick. This recipe tastes great served with wild rice or quinoa pilaf.

2 to 3 tablespoons high-quality butter or ghee

2 tablespoons coconut oil or avocado oil

½ pound large raw shrimp, shelled and deveined

3 or 4 garlic cloves, minced

¼ teaspoon red pepper flakes

Sea salt

Freshly ground black pepper

1 tablespoon freshly squeezed lemon juice

2 tablespoons finely chopped fresh parsley

PER SERVING Calories: 342; Total Fat: 27g; Saturated Fat: 20g; Sodium: 457mg; Carbohydrates: 2g; Fiber: 0g; Protein: 24g

1. In a large skillet over high heat, melt the butter and oil. Reduce to medium-high heat, and stir in the shrimp, garlic, and red pepper flakes. Add salt and pepper to taste.

2. Once the shrimp are all pink, after about 5 minutes, turn off the heat and toss with the lemon juice.

3. Sprinkle with a sprinkle of parsley.

STORAGE NOTE: This dish will keep, refrigerated, for 2 to 3 days.

RECIPE TIP: If you are avoiding shellfish, this dish is easily revised by using chicken chunks instead of shrimp.

Steamed Cod and Ginger over Wild Rice

Serves 2

PREP TIME: 20 MINUTES / COOK TIME: 10 MINUTES

This recipe presents itself as an elegant, fancy restaurant–style dish, but it's one of the simplest to prepare. Flaky, tangy, and very savory.

2 cups wild rice

3 tablespoons unpasteurized apple
    cider vinegar

2 tablespoons tamari or gluten-free
    soy sauce

2 tablespoons peeled and grated
    fresh ginger

2 (4-ounce) skinless cod fillets

Sea salt

Freshly ground black pepper

6 scallions, cut lengthwise into strings

PER SERVING Calories: 708; Total Fat: 3g;
Saturated Fat: 0g; Sodium: 1,120mg; Carbohydrates: 128g;
Fiber: 11g; Protein: 46g

1. Cook the wild rice per package directions. Set aside.

2. In a medium skillet over medium-high heat, combine the apple cider vinegar, tamari or soy sauce, and ginger.

3. Season both sides of the cod fillets with salt and pepper. Place the fillets in the skillet with the vinegar mixture. Bring to a boil, then reduce the heat. Cover and cook for 6 to 8 minutes, until the cod flakes easily with a fork. About 3 minutes before it's done, sprinkle the scallion strings over the fish and cook for the remaining 3 minutes

4. Serve the fish over wild rice.

STORAGE NOTE: This dish will keep, refrigerated, for 2 to 3 days.

RECIPE TIP: The rice can be made in advance and warmed for serving, However, the fish should be cooked to serve, as the soy can dry out the cod after refrigeration.

Fish Chowder

Serves 2

PREP TIME: 10 MINUTES / COOK TIME: 15 MINUTES

Quick and simple, clean, and easy to digest, this dish can be enjoyed for lunch, dinner, snack, and yes, even breakfast. I keep saying, don't overlook the good old-fashioned cup of soup for breakfast. For a more filling soup, pour in a can of unsweetened coconut milk.

2 tablespoons coconut or avocado oil

2 large carrots, peeled and roughly chopped

2 large celery stalks, roughly chopped

1 large onion, roughly chopped

Sea salt

Freshly ground black pepper

6 cups vegetable or chicken broth (homemade is great, but store-bought is fine)

2 white fish fillets (cod, mahi mahi, halibut), chopped into bite-size chunks

---

PER SERVING Calories: 392; Total Fat: 19g; Saturated Fat: 13g; Sodium: 1,230mg; Carbohydrates: 19g; Fiber: 4g; Protein: 36g

1. In a large saucepan over medium-high heat, heat the oil. Add the carrots, celery, onion, salt, and pepper, and cook, stirring occasionally, until the onion is shimmery and translucent, about 5 minutes.

2. Add the broth, and bring to a low simmer.

3. Add the fish chunks and cook for 3 minutes, or until the fish is tender but cooked through, and serve.

STORAGE NOTE: This dish will keep refrigerated for 2 to 3 days. It also freezes well.

RECIPE TIP: If you are avoiding fish, this dish is easily made with chicken chunks, or just keep it vegetarian and add more veggies, like diced potatoes, frozen peas, and corn. Spice it up with some cilantro at the end, or add a handful of spices, like garlic and chili flakes. This keeps well and makes a great wholesome lunch that won't weigh you down midday.

Mushroom and Asparagus "Risotto"

Serves 2

PREP TIME: 10 MINUTES / COOK TIME: 20 MINUTES

You can make this recipe over and over and differently every time, using vegetables like frozen peas, summer squash, or different types of mushrooms. This dish works as a main course or as a welcome addition to any main.

1 cauliflower head

3 tablespoons coconut oil, avocado oil, or high-quality butter or ghee, divided

Sea salt

Freshly ground black pepper

2 garlic cloves, grated or crushed

10 asparagus spears, trimmed and chopped into bite-size pieces

1 cup sliced cremini mushrooms

½ cup water

¼ cup unsweetened coconut milk

PER SERVING Calories: 365; Total Fat: 25g; Saturated Fat: 17g; Sodium: 375mg; Carbohydrates: 31g; Fiber: 14g; Protein: 13g

1. Cut the cauliflower into large florets. In a food processor, shred the cauliflower into "rice" with the shredding blade, or pulse with the chopping blade. Alternatively, carefully grate by hand on a box grater.

2. In a large skillet over medium heat, melt 1 tablespoon of oil. Add the cauliflower rice and sauté for 5 minutes. Season with salt and pepper to taste. Remove from the skillet and set aside.

3. In the skillet, heat the remaining 2 tablespoons of oil. Add the garlic and sauté for 2 minutes, being careful not to burn the garlic, then add the cauliflower rice, asparagus, mushrooms, and water. Cook for 3 minutes, then cover and cook for another 5 minutes.

4. Add the coconut milk, salt, and pepper, and continue to stir until the liquid is absorbed and the rice is creamy, 5 to 7 minutes.

5. Serve immediately.

STORAGE NOTE: This dish will keep, refrigerated, for 2 to 3 days.

Creamy Vegetable Alfredo

Serves 1

PREP TIME: 5 MINUTES / COOK TIME: 20 MINUTES

This Alfredo sauce is delicious enough to eat like a soup, but it's also amazing poured onto roasted vegetables or into a gluten-free pasta, and it is incredibly easy to digest.

1 cup salted water

1 cauliflower head, cut into florets

4 tablespoons coconut oil or avocado oil, divided

2 garlic cloves, minced

1 cup unsweetened coconut milk

1 cup vegetable broth

1 teaspoon Italian seasoning

Sea salt

Freshly ground black pepper

PER SERVING Calories: 1,226; Total Fat: 115g; Saturated Fat: 99g; Sodium: 1,208mg; Carbohydrates: 47g; Fiber: 20g; Protein: 22g

1. In a large saucepan with 1 cup of salted water, cover the cauliflower and steam in boiling water until tender and pierced with a fork easily, 5 to 7 minutes. Drain and set aside.

2. In the same saucepan, heat 1 tablespoon of oil and add the garlic. Sauté the garlic lightly, about 2 minutes. Do not brown.

3. Add the coconut milk. Once it begins to simmer, add the broth and Italian seasoning, and season with salt and pepper. Simmer for 5 minutes.

4. Transfer the ingredients to a blender. Carefully blend the ingredients until completely smooth (use caution when blending hot liquid) and serve.

STORAGE NOTE: This recipe is best enjoyed immediately.

RECIPE TIP: For a creamier consistency, add 1 cup of cashews to the blender.

Lettuce Cup Veggie Tacos

Serves 1

PREP TIME: 10 MINUTES / COOK TIME: 20 MINUTES

This taco "meat" is so versatile. You can use it alone as a dip with blue corn chips, fry it with some quinoa or wild rice, or serve on some lettuce cups and enjoy a meatless taco Tuesday.

1 cup lentils

Sea salt

Freshly ground black pepper

½ cup chopped walnuts

2 garlic cloves, peeled

3 canned chipotle chiles with sauce

½ tablespoon coconut oil or avocado oil

Romaine hearts, for serving

PER SERVING Calories: 868; Total Fat: 20g; Saturated Fat: 7g; Sodium: 577mg; Carbohydrates: 122g; Fiber: 63g; Protein: 55g

1. In a medium saucepan of water, bring the lentils to a low boil until tender, about 20 minutes. Add salt and pepper to taste.

2. Transfer the lentils and the walnuts, garlic, chiles, and oil to a blender or food processor, and pulse into a ground-beef-like texture.

3. Serve the mixture on romaine lettuce cups.

STORAGE NOTE: The "taco meat" keeps, refrigerated, for 1 week.

RECIPE TIP: Lentils can be a bit hard to digest for some, so make sure they are cooked all the way through. I love to set up a taco bar with these and use condiments like guacamole, fermented veggies, diced tomatoes, or even a little feta cheese.

Sweet Potato and Spinach Comfort Bowl with Pumpkin Seeds

Serves 1

PREP TIME: 5 MINUTES / COOK TIME: 15 MINUTES

If a cozy bathrobe and slippers could be a dish, it would be this dish. Truly the epitome of comfort food, this bowl is filling, warm, and delicious with a little crunch.

1 large garnet yam or sweet potato, peeled and sliced into bite-size pieces

1 tablespoon coconut oil or avocado oil

1 (5-ounce) bag loose-leaf spinach

1 cup cooked quinoa

½ avocado, sliced

¼ cup Toasted Pumpkin Seeds with Sea Salt (page 130)

PER SERVING Calories: 815; Total Fat: 47g; Saturated Fat: 17g; Sodium: 210mg; Carbohydrates: 84g; Fiber: 19g; Protein: 24g

1. In a medium saucepan, cover the yam pieces with water and bring to a boil. Cook until you can pierce the yam pieces with a knife easily but they do not fall apart, 7 to 10 minutes. Drain and set aside.

2. In a large skillet over medium heat, heat the oil. Add the spinach and turn the leaves with tongs until wilted, about 3 minutes. Set aside.

3. In a large serving bowl, add the quinoa, then the spinach, then the yams. Top with the avocado slices and Toasted Pumpkin Seeds and serve.

STORAGE NOTE: This dish keeps, refrigerated, for 1 week. Add the avocado and pumpkin seeds just before serving.

RECIPE TIP: Making the quinoa, toasting the pumpkin seeds, and cooking the yams in advance make this one of the easiest dishes when you are in a hurry or too tired to make dinner. I love to add leftover Squash and Ginger Soup (page 87) as a dressing for this dish.

Broccoli-Fennel Soup

Serves 1

PREP TIME: 5 MINUTES / COOK TIME: 20 MINUTES

This is the easiest soup you can make. It's incredibly elegant with the subtle fennel flavor and is a wonderful soup starter to impress dinner party guests.

2 tablespoons coconut oil or avocado oil

1 large onion, roughly chopped

1 large broccoli head, chopped into florets

1 fennel bulb, roughly chopped

6 cups vegetable broth

Sea salt

Freshly ground black pepper

PER SERVING Calories: 581; Total Fat: 29g;
Saturated Fat: 24g; Sodium: 2,344mg; Carbohydrates: 73g;
Fiber: 26g; Protein: 14g

1. In a large stock pot or saucepan over medium-high heat, heat the oil. Sauté the onion until shimmery and translucent, about 5 minutes.

2. Add the broccoli and fennel, and sauté until both are tender but still green and vibrant, 5 to 7 minutes.

3. Add the vegetable broth, and bring to a low boil.

4. Turn off the heat. Gently pour the soup into a blender, and carefully liquefy until smooth (use caution when blending hot liquid). Season with salt and pepper and serve.

STORAGE NOTE: This soup is best eaten immediately, but it will freeze well. Adding a can of coconut milk will make this dish more silky and creamy.

RECIPE TIP: Soaking the rice overnight can help with digestion. If a vegetarian diet isn't an issue, you can use chicken broth instead of vegetable.

Squash and Ginger Soup

Serves 2

PREP TIME: 15 MINUTES / COOK TIME: 20 MINUTES

I make this soup as a starter every single Thanksgiving, and it's a huge hit. I also use this soup as a sauce on roasted vegetables or to upgrade quinoa, or even as a dressing for the Sweet Potato and Spinach Comfort Bowl with Pumpkin Seeds (page 85).

2 tablespoons coconut oil or avocado oil

1 large onion, roughly chopped

1 large butternut squash, peeled, seeded, and chopped into medium-size chunks

½-inch piece ginger, peeled and grated

Sea salt

Freshly ground black pepper

6 cups vegetable broth

1 (13.5-ounce) can unsweetened coconut milk

PER SERVING Calories: 790; Total Fat: 60g; Saturated Fat: 52g; Sodium: 1,253mg; Carbohydrates: 68g; Fiber: 16g; Protein: 9g

1. In large stock pot or saucepan over medium-high heat, heat the oil. Sauté the onion until shimmery and translucent, about 5 minutes.

2. Add the squash and ginger, season with salt and pepper, and cook for 3 minutes, stirring to coat.

3. Add the broth and coconut milk, and bring to a low simmer. Simmer until the squash can be pierced with a knife easily, 7 to 10 minutes.

4. Carefully add the ingredients to a blender, liquefy (use caution when blending hot liquid), and serve.

STORAGE NOTE: This soup is best eaten immediately, but you can freeze individual containers and take it on the go for lunch or a snack.

RECIPE TIP: I like to throw in chopped carrots with the onions for an even more layered flavor. You can also play with spices like nutmeg and cloves for a rich fall flavor. If a vegetarian diet isn't an issue, you can use chicken broth instead of vegetable broth.

One-Pot Zucchini Noodle Miso

Serves 1

PREP TIME: 10 MINUTES / COOK TIME: 20 MINUTES

You will need to buy a vegetable spiralizer, which these days can be purchased anywhere. In fact, I bought mine at the 99¢ store; no need for anything fancy. And miso paste is a great addition to your diet; it's easy on digestion and always fermented. This recipe calls for brown rice miso, but you can use any Miso Master products. They are all high quality and easy to find. A warm cup of miso all by itself is great for breakfast, lunch, dinner, or snack. If you use other brands, be careful of the added ingredients, which often include MSG, which is not a friend to a healthy gut.

1 tablespoon coconut oil or avocado oil

1 large zucchini, spiraled

10 asparagus spears, cut in half, then halved lengthwise

1 to 2 teaspoons hot sauce or 1 teaspoon chili flakes

Sea salt

Freshly ground black pepper

2 tablespoons brown rice miso paste

2 cups water

2 or 3 scallions, minced

PER SERVING Calories: 293; Total Fat: 15g; Saturated Fat: 12g; Sodium: 1,852mg; Carbohydrates: 36g; Fiber: 10g; Protein: 10g

1. In a large saucepan over medium-high heat, heat the oil. Add the zucchini noodles and asparagus, and sauté until well coated.

2. Add the hot sauce and salt and pepper to taste, and sauté for 1 minute more.

3. Add the miso paste and heat until it begins to melt. Add the water, and bring to a low simmer. Turn off the heat and serve garnished with scallions.

STORAGE NOTE: Best eaten immediately, this can be refrigerated for up to 1 week. It freezes well, too.

RECIPE TIP: You can really take the flavor up a notch in this recipe by adding sautéed garlic, sesame seeds, or a soft-boiled egg!

Stuffed Baked Sweet (or Russet) Potatoes

Serves 2

PREP TIME: 10 MINUTES / COOK TIME: 45 MINUTES

Sweet potatoes or yams are the most overlooked food for versatility. Other garnishes for this recipe include scallions, chopped tomatoes, fermented veggies, and how about some of that vegan taco meat you made for the Lettuce Cup Veggie Tacos recipe (page 84)? And a really nice surprise is a breakfast yam filled with almond butter, cinnamon, and bananas. Delicious.

2 large garnet yams, scrubbed well (do not poke holes in them)

1 tablespoon coconut oil

1 cup cooked or canned black beans

Sea salt

Freshly ground black pepper

½ cup plain Greek yogurt

Juice of 1 lime

½ cup fresh cilantro, chopped

PER SERVING Calories: 479; Total Fat: 8g; Saturated Fat: 7g; Sodium: 165mg; Carbohydrates: 86g; Fiber: 17g; Protein: 17g

1. Preheat the oven to 420°F.

2. On a large, rimmed baking sheet, bake the yams until easily pierced with a knife, about 45 minutes, depending on size.

3. Meanwhile, in a medium saucepan over medium heat, melt the coconut oil. Add the beans and heat until warm. Add salt and pepper to taste, and set aside.

4. In a small bowl, mix together the Greek yogurt with the lime juice.

5. Carefully slice open the yams with a knife, top with the beans, lime cream sauce, and cilantro, and serve.

STORAGE NOTE: Yams keep well once baked; simply refrigerate in an airtight container for lunch tomorrow, or up to 1 week. But once it's stuffed, it's best eaten immediately.

RECIPE TIP: I love russet potatoes in place of yams for the amazing prebiotic content in large, starchy potatoes—great for the gut. Canned beans can be used here, though cooked are preferable. Canned black beans can be difficult to digest.

Brown Rice Pasta with Arugula, Feta, and Pine Nuts

Serves 4

PREP TIME: 5 MINUTES / COOK TIME: 20 MINUTES

Guilt-free pasta! This bright and light pasta makes a great lunch; add a grilled piece of salmon to make it a delicious complete meal. The ease and crunchy pine nuts make this dish a real favorite.

1 pound gluten-free pasta, cooked al dente

1 (5-ounce) bag loose leaf arugula, roughly chopped

1 cup finely chopped fresh basil

2 tablespoons avocado or olive oil, plus more if needed

1 cup crumbled feta cheese

½ cup pine nuts or toasted almond slivers (see page 31)

Sea salt

Freshly ground black pepper

PER SERVING Calories: 683; Total Fat: 29g; Saturated Fat: 8g; Sodium: 487mg; Carbohydrates: 93g; Fiber: 3g; Protein: 17g

1. In a large bowl, toss the warm pasta with the arugula, basil, and oil to coat well.

2. Sprinkle with the feta, nuts, salt, pepper, and extra oil if needed.

3. Serve warm.

STORAGE NOTE: This dish is best eaten immediately.

RECIPE TIP: You can replace the nuts with more veggies or pomegranate seeds for a crunch.

Vegan Gluten-Free Falafel

Serves 2

PREP TIME: 10 MINUTES, PLUS 30 MINUTES TO CHILL / COOK TIME: 10 MINUTES

This vegan treat is loaded with Mediterranean flavor. And it's so easy to make your own oat flour to ensure they are gluten-free. These can be served with Greek Quinoa Tabbouleh (page 70) or wrapped in lettuce leaves.

½ cup gluten-free oats

1 can organic chickpeas, drained and rinsed, liquid reserved

½ cup onion, chopped

¼ cup chopped fresh parsley

2 tablespoons avocado oil, divided

1 teaspoon ground cumin

Sea salt

Freshly ground black pepper

PER SERVING Calories: 498; Total Fat: 18g; Saturated Fat: 12g; Sodium: 843mg; Carbohydrates: 72g; Fiber: 14g; Protein: 15g

1. In a blender or food processor, blend or pulse the oats until pulverized into a flour consistency.

2. Add the chickpeas, onion, parsley, 1 tablespoon of oil, and the cumin to the food processor, and season with salt and pepper. Pulse until completely combined into a dough. Add a little chickpea liquid if needed to help form the dough.

3. Refrigerate the dough for 30 minutes.

4. Form the dough into small patties, about half the size of your palm.

5. In a large skillet over medium-high heat, heat the remaining tablespoon of oil.

6. Cook the patties for 3 to 5 minutes on each side, or until crispy.

7. Drain on paper towels. Serve with a side of Greek Quinoa Tabbouleh (page 70).

STORAGE NOTE: Falafel can be refrigerated for 5 days in an airtight container. They also freeze well.

RECIPE TIP: Chickpeas can be hard for some people to digest. If this applies to you, avoiding all beans during the diet is best.

Honey-Garlic Pork Medallions and Broccoli, page 104

Simple Healthy Meat Loaf

Serves 4

PREP TIME: 10 MINUTES / COOK TIME: 1 HOUR

Something I like to remember when it comes to gut health and our digestive system is: Our guts like things to be easy! The more ingredients, the more complex a dish, the more challenging it is to digest. This dish is full of flavor—the comforting, hearty deliciousness of meat loaf without the sugar, gluten, and bloating.

QUICK PREP

1 pound 90% lean grass-fed ground beef

½ cup diced onion

1 free-range egg

¼ cup Homemade Ketchup (page 132), plus more for topping

1 cup grated Parmesan cheese

½ teaspoon sea salt

½ teaspoon freshly ground black pepper

Healthy BBQ Sauce (page 137, optional), plus more for topping

PER SERVING Calories: 327; Total Fat: 18g; Saturated Fat: 9g; Sodium: 789mg; Carbohydrates: 6g; Fiber: 0g; Protein: 35g

1. Preheat the oven to 350°F.

2. Oil a loaf pan, or line with parchment or aluminum foil.

3. In a large bowl, use your hands to combine the beef, onion, egg, Homemade Ketchup, cheese, salt, pepper, and Healthy BBQ Sauce well.

4. Transfer the mixture to the loaf pan, and shape to fit.

5. If desired, spread some BBQ sauce or extra homemade ketchup on top of the loaf.

6. Bake for 45 minutes to 1 hour, depending on desired doneness.

7. Serve with a side of Buttery Yams and Walnuts (page 74).

STORAGE NOTE: This dish will keep, refrigerated, for 1 week.

RECIPE TIP: Make this meat loaf and cut it into individual servings and freeze. It's so easy to thaw a piece out at night and take it for lunch the next day.

Pan-Fried Bison Patty with Baby Spinach and Pink Cabbage Kimchi

Serves 1

PREP TIME: 5 MINUTES / COOK TIME: 10 MINUTES

This is my version of fast food. Healthy, delicious, filled with flavor, and quick. When cooked properly and eaten with fermented vegetables, this gut-friendly meal is sure to be a weekly go-to.

¼ **pound ground bison**

Sea salt

Freshly ground black pepper

2 tablespoons coconut oil or avocado oil, divided

2 cups baby spinach

1 cup Pink Cabbage and Garlic Fermented Kimchi (page 62)

PER SERVING Calories: 498; Total Fat: 38g; Saturated Fat: 29g; Sodium: 1,056mg; Carbohydrates: 6g; Fiber: 1g; Protein: 25g

1. Season the bison with salt and pepper, and use your hands to mix and form a patty the size of the palm of your hand.

2. In a large skillet over medium-high heat, heat 1 tablespoon of oil. Add the spinach and sauté until cooked down, about 3 minutes. Set on a serving plate.

3. In the skillet, heat the remaining table-spoon of oil. Place the patty in the skillet. Cook for 3 minutes on each side, being careful not to overcook.

4. Place the patty on top of the spinach, and serve with a side of Pink Cabbage and Garlic Fermented Kimchi.

5. Sprinkle with salt and pepper to taste.

STORAGE NOTE: This dish is best eaten immediately. You can premake and form these patties then freeze or refrigerate for later.

RECIPE TIP: You can turn up the flavor by adding chopped onions, garlic, and even some Homemade Ketchup (page 132), Healthy BBQ Sauce (page 137), or some Fermented Thai Chile Sauce (page 136).

Rosemary and Garlic Roast Pork Loin

Serves 4

PREP TIME: 10 MINUTES / COOK TIME: 30 MINUTES, PLUS 10 MINUTES TO REST

This rustic, classic dish can be prepared either for a holiday dinner or simply for lunches for the week. Serve with a large salad or any of the vegetable sides in this book. It also makes a perfect Sunday roast.

2 garlic cloves, minced

2 tablespoons Dijon mustard

1 tablespoon fresh rosemary, minced finely

1 tablespoon avocado oil, plus more for greasing the pan

2 teaspoons sea salt

¼ teaspoon freshly ground black pepper

1 (2-pound) boneless pork loin

PER SERVING Calories: 341; Total Fat: 14g; Saturated Fat: 4g; Sodium: 1,145mg; Carbohydrates: 2g; Fiber: 1g; Protein: 49g

1. Preheat the oven to 400°F.

2. In a small bowl, mix the garlic, mustard, rosemary, oil, salt, and pepper.

3. Grease a baking dish with avocado oil.

4. Place the pork loin in the dish, and cover all over with the herb mixture.

5. Cover with aluminum foil and bake for 20 minutes. Uncover and cook for an additional 10 minutes, or until the internal temperature reaches 140 to 150°F. The meat should be pale and mostly white with mostly clear juices.

6. Remove from the oven and let sit, covered, for 10 minutes before serving.

STORAGE NOTE: You can freeze pork loin in segments and thaw out overnight for lunch or dinner the next evening. You can refrigerate this dish for 3 to 4 days.

RECIPE TIP: Cooking this dish in advance is a great way to prepare lunches for the week.

Lamb Bites, Roasted Root Vegetables, and Homemade Pesto

Serves 2

PREP TIME: 10 MINUTES / COOK TIME: 30 MINUTES

Quick and easy is the name of the game here, but there's so much more. The root vegetables are grounding, with loads of satisfying fiber and minerals. The lamb is a complete protein that is easier on digestion than beef. And the pesto upgrades the flavor in this dish beyond expectation.

2 parsnips, peeled and chopped into bite-size pieces

2 yellow beets, peeled and chopped into bite-size pieces

2 large sweet onions, roughly chopped

2 tablespoons avocado oil or coconut oil, divided

Sea salt

Freshly ground black pepper

8 to 12 ounces lamb shoulder, cut into bite-size pieces

1 cup Vegan Pesto (page 138)

PER SERVING Calories: 692; Total Fat: 56g; Saturated Fat: 10g; Sodium: 165mg; Carbohydrates: 26g; Fiber: 5g; Protein: 20g

1. Preheat the oven to 420°F.

2. On a large, rimmed baking sheet, toss the parsnips, beets, and onions in 1 tablespoon of oil, and sprinkle with salt and pepper to taste.

3. Roast the vegetables for 20 to 30 minutes, or until you can pierce the beets easily with a knife and they don't fall apart.

4. Meanwhile, heat a large skillet over medium-high heat and add the remaining tablespoon of oil. When the oil is hot and shimmery, add the lamb bites, salt and pepper to taste, and cook, stirring often. Sauté for 4 to 5 minutes. They should be slightly browning but still tender and ideally medium-rare.

Continued

Lamb Bites, Roasted Root Vegetables, and Homemade Pesto *Continued*

5. Transfer the lamb to a large bowl.

6. When the vegetables are cooked, add them to the bowl with the lamb and mix together. Add the Vegan Pesto, mix well, and serve.

STORAGE NOTE: This dish keeps refrigerated for up to 1 week. It freezes well and makes for great lunches.

RECIPE TIP: For vegetarians, omit the lamb and add more veggies, cauliflower, broccoli, and potatoes. The pesto makes this dish amazingly tasty, and subbing any protein will work in this recipe.

Garlic and Herb Roasted Lamb Chops with Tzatziki Sauce

Serves 2

PREP TIME: 15 MINUTES, PLUS 1 HOUR TO MARINATE /
COOK TIME: 10 MINUTES, PLUS 10 MINUTES TO STAND

I've always found lamb chops to be elegant and fancy, so I couldn't believe how easy they were to prepare. They go with just about any cooked vegetable side or salad, and dress up any dinner party. Lamb is dense in just about every vitamin and mineral, and high in protein and B12.

4 garlic cloves, crushed

1 tablespoon fresh rosemary, minced

1 tablespoon fresh parsley, minced

1 tablespoon fresh oregano, minced

2 tablespoons avocado oil, plus more for greasing the pan

2 tablespoons high-quality butter, softened

Sea salt

Freshly ground black pepper

6 lamb loin chops

½ cup Tzatziki (page 139)

PER SERVING Calories: 561; Total Fat: 52g; Saturated Fat: 30g; Sodium: 267mg; Carbohydrates: 5g; Fiber: 2g; Protein: 20g

1. Preheat the oven to 400°F.

2. In a small bowl, whisk together the garlic, rosemary, parsley, oregano, oil, butter, salt, and pepper.

3. Place the lamb chops in a shallow baking dish and rub with the marinade. Cover and marinate in the refrigerator for 1 hour, or up to overnight for more depth of flavor.

4. When done marinating, bring the chops to room temperature.

5. Liberally oil a rimmed baking sheet or large cast iron skillet. Place the chops on the baking sheet.

Continued

Garlic and Herb Roasted Lamb Chops with Tzatziki Sauce *Continued*

6. Roast for about 10 minutes (this can vary depending on thickness of chops; don't overcook). The meat is done when pale and mostly white with mostly clear juices.

7. Remove from the oven and let stand away from the heat, covered with foil, for 10 minutes before serving.

8. Serve 3 chops on each plate, each topped with ¼ cup Tzatziki. Pair with an arugula salad and Raspberry Vinaigrette (page 135).

STORAGE NOTE: These chops can keep, refrigerated, for about 4 days. They are great as a cold lunch paired with a side salad.

RECIPE TIP: The longer these chops sit in the marinade, the more delicious they become. Don't be shy with herbs; thyme, herbs de Provence, mustard—all go a long way when flavoring this dish. You can use dried herbs, but fresh ones contain more nutrients and flavor, and their stems are filled with beneficial bacteria.

Meatballs and Spaghetti Squash

Serves 2

PREP TIME: 15 MINUTES / COOK TIME: 1 HOUR

My husband is a New Jersey Italian through and through. This dish saved us when he was doing this diet, and it satisfied his craving for spaghetti and meatballs.

2 tablespoons avocado oil, divided, plus more for greasing the pan

1 large spaghetti squash, washed, halved lengthwise, and seeded

½ pound lean ground beef

1 free-range egg

2 tablespoons Italian seasoning, divided

3 garlic cloves, minced

Sea salt

Freshly ground black pepper

1 teaspoon olive oil

PER SERVING Calories: 469: Total Fat: 32g; Saturated Fat: 7g; Sodium: 266mg; Carbohydrates: 21g; Fiber: 0g; Protein: 28g

1. Preheat the oven to 420°F.

2. Grease a large, rimmed baking sheet. Bake the squash, flat-side down on the baking sheet, until it pierces easily with a knife, 30 to 45 minutes, depending on size. Set aside.

3. Reduce the oven temperature to 350°F.

4. In a large bowl, mix together the ground beef, egg, 1 tablespoon of avocado oil, 1 tablespoon of Italian seasoning, and garlic. Mix together well and form into balls the size of golf balls.

5. Oil the baking sheet again or line with parchment paper. Spread the meat-balls on top without crowding, and bake for 20 minutes, or until cooked through. Set aside.

Continued

Meatballs and Spaghetti Squash *Continued*

6. When the squash has cooled enough to handle, using a fork, fluff the inside of the squash into a spaghetti-like consistency. Scoop the strands into a large bowl and set aside until cool. Add the remaining tablespoon of avocado oil and the remaining tablespoon of Italian seasoning, and mix well.

7. Serve a heaping portion of squash onto a plate, and place a few meatballs on top. Drizzle with the olive oil, season with salt and pepper, and serve.

STORAGE NOTE: These meatballs freeze well, or you can store them in the refrigerator for up to a week. The squash is best eaten immediately.

RECIPE TIP: I make these meatballs on a Sunday and then refrigerate them, so they are all ready to bake during the week, easy as can be. A simple addition is to heat a can of crushed tomatoes and sautéed garlic, and voilà—you have a sauce.

Fajita Beef Bowl

Serves 2

PREP TIME: 5 MINUTES / COOK TIME: 15 MINUTES

For some reason kids love fajitas, and you could get away with changing out the protein on this over and over. Try shrimp, chicken, steak, and sometimes just vegetables. Add guacamole and blue corn chips with a little salsa on the side, and make a fiesta out of it.

2 tablespoons avocado oil

1 pound skirt steak or flank steak, cut into strips and brought to room temperature

1 large sweet onion, cut into strips

2 garlic cloves, minced

1 bell pepper, cored, seeded, and cut into strips

1 tablespoon fajita seasoning or southwest seasoning

Sea salt

Freshly ground black pepper

PER SERVING Calories: 538; Total Fat: 31g; Saturated Fat: 9g; Sodium: 514mg; Carbohydrates: 16g; Fiber: 3g; Protein: 47g

1. In a large skillet over medium-high heat, heat the oil. Cook the steak strips until still a bit pink inside, 3 to 4 minutes. Set aside.

2. Add the onion and garlic to the pan, and sauté until the onion is translucent and shiny, about 5 minutes.

3. Add the bell pepper and fajita seasoning, season with salt and pepper, and cook for about 3 minutes. The peppers should still be crunchy and vibrant and not limp.

4. Return the steak strips to the skillet and turn off the heat. Mix all the ingredients well.

5. Serve with a side of cooked brown rice and Pink Cabbage and Garlic Fermented Kimchi (page 62).

STORAGE NOTE: Refrigerate for up to 4 days.

RECIPE TIP: If you are avoiding nightshades for any reason, you can substitute broccoli for the bell pepper.

Honey-Garlic Pork Medallions and Broccoli

Serves 4

PREP TIME: 10 MINUTES / COOK TIME: 15 MINUTES

This pork tenderloin dish is a simple enough meal to make in advance for a whole week of lunches, or fancy enough to serve at a holiday dinner. Upgrade the flavor even further with some Dijon mustard and Fermented Thai Chile Sauce (page 136) in the vinegar sauce.

½ cup balsamic vinegar

2 tablespoons raw honey

4 garlic cloves, finely minced

Sea salt

Freshly ground black pepper

1 tablespoon avocado oil

1 (1-pound) pork tenderloin, cut into 1-inch-thick medallions

1 broccoli head, cut into florets

PER SERVING Calories: 224; Total Fat: 7g;
Saturated Fat: 2g; Sodium: 391mg; Carbohydrates: 16g;
Fiber: 3g; Protein: 24g

1. Preheat the oven to 350°F.

2. In a small bowl, whisk together the vinegar, honey, and garlic, and season with salt and pepper.

3. In a large skillet over medium-high heat, heat the oil. Add the medallions, being careful not to crowd the pan.

4. Sear for 1 minute on each side, then transfer the medallions to a baking dish. Pour the vinegar mixture over the medallions.

5. Add the broccoli florets to the skillet, and sauté just enough to coat with the remaining juices in the pan, then transfer the broccoli and juices into the baking dish with the pork.

6. Cover the dish with a lid or foil, and roast for 10 minutes. Remove from the oven and serve.

STORAGE NOTE: Refrigerate this dish in an airtight container for up to 1 week.

RECIPE TIP: You can use a thermometer and remove from the oven once the pork reaches 140°F.

Garlic-Baked Chicken, Sesame-Sautéed Cabbage, and Baked Potato Fries

Serves 2

PREP TIME: 5 MINUTES / COOK TIME: 45 MINUTES

This wholesome recipe will fill you up but not bloat you. Cooking the cabbage through until tender helps extract all the wonderful nutrients that cabbage can offer, and the sesame oil gives it a wonderful toasty Asian flavor. The potato fries were a last-minute addition for me one day, and it made the whole meal, so I've made them part of this recipe.

1 precut half chicken

2 tablespoons avocado oil or coconut oil, divided

Sea salt

Freshly ground black pepper

4 garlic cloves, finely minced

1 cabbage head, cored and thinly sliced

1 teaspoon toasted sesame oil

Baked Russet French Fries with Parsley and Garlic (page 73)

PER SERVING Calories: 585; Total Fat: 29g; Saturated Fat: 16g; Sodium: 881mg; Carbohydrates: 49g; Fiber: 13g; Protein: 29g

1. Preheat the oven to 375°F.

2. In a three-quarter sheet pan, pat the half chicken dry. Drizzle with 1 tablespoon of oil, add salt and pepper to both sides, and sprinkle with about half of the minced garlic.

3. Roast for 35 to 40 minutes, or until the chicken is cooked through and the juice in the dish is no longer pink.

4. Meanwhile, in a large skillet over medium heat, heat the remaining tablespoon of oil until shimmering.

5. Add the other half of the garlic and the sliced cabbage, sprinkle with salt and pepper, and sauté until shimmery and tender, about 5 minutes.

Continued

Garlic-Baked Chicken, Sesame-Sautéed Cabbage, and Baked Potato Fries *Continued*

6. In a large bowl, drizzle the cabbage with the toasted sesame oil and mix well.

7. Serve the cabbage with the roasted chicken and Baked Russet French Fries with Parsley and Garlic.

STORAGE NOTE: This dish keeps refrigerated for up to 1 week. It also freezes well.

RECIPE TIP: You can double this recipe and literally have dinners for a week. Also, make the French fries in advance, and heat in a warm oven to serve with this dish.

Chicken-Avocado-Lime Soup
Serves 1
PREP TIME: 5 MINUTES / COOK TIME: 20 MINUTES

This is my go-to soup when anyone is feeling under the weather. If I don't have time to roast a chicken to make it the old-fashioned way, this is just as good. Add some frozen corn to upgrade the flavor. This recipe is so easy on the digestive system, and soothing and healing to the tummy.

1 tablespoon coconut oil or avocado oil

2 boneless chicken breasts, cut into bite-size pieces

Sea salt

Freshly ground black pepper

2 garlic cloves, minced

4 cups chicken broth or 2 cans full-fat unsweetened coconut milk

1 tablespoon freshly squeezed lime juice

1 large ripe avocado, peeled, pitted, and cut into thin slices

½ cup chopped fresh cilantro

PER SERVING Calories: 728; Total Fat: 43g; Saturated Fat: 16g; Sodium: 1,543mg; Carbohydrates: 25g; Fiber: 12g; Protein: 64g

1. In a large stockpot over medium-high heat, heat the oil.

2. When hot, add the chicken, salt and pepper to taste, and sear until browned all over, about 3 minutes on each side.

3. Add the garlic and sauté for 1 minute, then add the chicken broth.

4. Bring to a low boil for 10 minutes.

5. Spoon into a serving bowl, add the lime juice, avocado, and cilantro, and serve.

STORAGE NOTE: This soup is best eaten immediately. It freezes well and will keep, refrigerated, for up to 1 week. However, always cut and add the avocado right before serving, not before storing.

RECIPE TIP: This is a great soup as a starter, snack, and lunch. Premake the chicken and have it handy in a pinch. You can boost the nutrition in this recipe by adding 2 cups of roughly chopped baby red potatoes.

Curried Chicken Legs with Carrots

Serves 4

PREP TIME: 10 MINUTES / COOK TIME: 35 MINUTES

Chicken legs are nature's fast food. You can take this dish to a picnic and eat the chicken cold. Upgrade the nutrition by adding more vegetables—broccoli, cauliflower, and bell peppers are great additions.

8 chicken legs

Sea salt

Freshly ground black pepper

2 tablespoons coconut oil, divided

2 medium onions, thinly sliced

2 tablespoons yellow curry

3 large carrots, peeled and cut into discs, then into crescents

2 cans unsweetened coconut milk

Precooked rice (optional)

PER SERVING Calories: 939; Total Fat: 80g; Saturated Fat: 52g; Sodium: 486mg; Carbohydrates: 22g; Fiber: 8g; Protein: 42g

1. Pat the chicken dry, and salt and pepper both sides.

2. In a Dutch oven or large pot with a lid over medium-high heat, heat 1 tablespoon of oil.

3. When the oil is hot, add the chicken and brown on all sides, about 3 minutes. Transfer to a plate and set aside.

4. Add the remaining tablespoon of oil to the pot over medium-high heat. Add the onions and curry, sprinkle with salt and pepper, and sauté until translucent, about 3 minutes.

5. Add the carrots to the pot. Return the chicken to the pot, along with the coconut milk.

6. Add enough water just to cover the chicken and vegetables.

7. Cover and bring to a low simmer for 30 minutes.

8. Serve over warm, precooked rice (if desired).

STORAGE NOTE: This dish keeps, refrigerated, up to 1 week. It also freezes well.

Spring Chicken and Peas

Serves 1

PREP TIME: 5 MINUTES / COOK TIME: 15 MINUTES

This simple dish can be served hot or cold. I have also mixed in a couple of tablespoons of my Homemade Vegan Mayo (page 133) and turned this into a classic chicken salad—light, fresh, and delicious.

2 tablespoons avocado oil, divided

1 (5-ounce) bag loose leaf spinach

2 boneless chicken breasts, cut into bite-size pieces

Sea salt

Freshly ground black pepper

2 cups frozen peas

1 cup pine nuts

½ cup Parmesan cheese

PER SERVING Calories: 1,544; Total Fat: 110g; Saturated Fat: 22g; Sodium: 1,016mg; Carbohydrates: 42g; Fiber: 14g; Protein: 111g

1. In a large skillet over medium-high heat, heat 1 tablespoon of oil.

2. Add the spinach, and sauté until cooked down, about 3 minutes. It should not be completely wilted. Transfer to a plate and set aside.

3. Add the chicken, sprinkle with salt and pepper, and sauté until cooked through, about 4 minutes on each side.

4. Add the frozen peas and spinach, and cook for 5 more minutes.

5. Sprinkle with the pine nuts and Parmesan and serve.

STORAGE NOTE: This dish freezes well and will keep in the refrigerator for up to 1 week.

RECIPE TIP: For a more filling meal, cook up some gluten-free pasta and mix it all together.

Ranch Chicken and Veggie Skillet

Serves 2

PREP TIME: 10 MINUTES / COOK TIME: 20 MINUTES

This skillet is easy and filled with every single nutrient needed in the human body. The herbs are densely nutritious and easy to digest. Adding vegetable or bone broth instead of water adds even more nutrition. Serve over quinoa or brown rice for a more filling meal.

1 tablespoon avocado or coconut oil

4 boneless chicken thighs, cut into
    bite-size pieces

Sea salt

½ cup water

1 broccoli head, cut into florets

½ cup minced chives

½ cup minced fresh parsley

½ cup minced fresh dill

Freshly ground black pepper

PER SERVING Calories: 520; Total Fat: 34g;
Saturated Fat: 14g; Sodium: 338mg; Carbohydrates: 20g;
Fiber: 7g; Protein: 38g

1. In a large skillet with a lid over high heat, heat the oil until hot and shimmery. Carefully add the chicken, and sprinkle with salt, stirring quickly (you don't want the chicken to stick to the pan).

2. Once the chicken bites are well coated with oil and browning a little, after 3 to 4 minutes, transfer to a plate and set aside.

3. Lower the heat to medium-low, and add the water. Use a spatula to scrape up the browned bits of chicken, stirring around to make a brown liquid.

4. Add the broccoli, chives, parsley, and dill, and season with salt and pepper, stirring until well coated.

5. Return the chicken to the skillet. Cover and cook for 10 minutes, or until the chicken is cooked through and the broccoli is tender but still vibrant green, and serve.

STORAGE NOTE: This dish will refrigerate well in an airtight container for up to 1 week, and it freezes well, too.

RECIPE TIP: Premake this dish and use it for multiple meals throughout the week.

One-Pot Lemon-Herbed Chicken and Quinoa

Serves 4

PREP TIME: 5 MINUTES / COOK TIME: 25 MINUTES

I wish life was "one pot." So easy. Italian seasoning is a convenient way to get loads of flavor in this dish, but you can also add fresh herbs like rosemary, oregano, thyme, and even mint. Get fancy and garnish it with lemons. I have also used chicken thighs in this recipe for a much moister outcome.

2 tablespoons avocado oil, divided

4 chicken breasts

3 teaspoons Italian seasoning

Sea salt

Freshly ground black pepper

1 cup uncooked quinoa

3 tablespoons freshly squeezed lemon juice

2 cups chicken or vegetable broth

PER SERVING Calories: 377; Total Fat: 14g; Saturated Fat: 2g; Sodium: 503mg; Carbohydrates: 28g; Fiber: 3g; Protein: 32g

1. In a large skillet with a lid over medium-high heat, heat 1 tablespoon of oil.

2. Add the chicken, sprinkle with 1 teaspoon of Italian seasoning, and season with salt and pepper. Cook for about 2 minutes on each side, not until cooked through. Remove from the heat, and set aside on a plate.

3. Add the remaining tablespoon of oil and the quinoa to the hot skillet, stirring and coating well.

4. Return the chicken to the skillet, along with the lemon juice, remaining 2 teaspoons of Italian seasoning, and broth, cover, and reduce the heat to a simmer.

Continued

5. Simmer on low for 20 minutes, or until the liquid has absorbed and the quinoa is fluffy, and serve.

STORAGE NOTE: This dish stores in the refrigerator for up to 1 week.

RECIPE TIP: I have made this dish with lamb, fish, and even chopped sirloin. If you are a vegetarian, just replace the chicken with mixed veggies and reduce simmering time.

Chicken-Coconut Curry over Brown Rice

Serves 2

PREP TIME: 10 MINUTES / COOK TIME: 20 MINUTES

Curry is normally a dish that takes hours and many ingredients to get the depth of flavor for which it is famous; however, the complexity of most curries can also be hard on the gut. Here we manage to achieve that desired flavor with this simple, clean recipe.

1 tablespoon coconut oil

1 onion, diced

1 red bell pepper, diced

3 large skinless, boneless chicken breasts, chopped into bite-size pieces

2 (13.5-ounce) cans unsweetened coconut milk

4 tablespoons yellow curry powder

Sea salt

Freshly ground black pepper

2 cups precooked brown rice

PER SERVING Calories: 1,887; Total Fat: 107g; Saturated Fat: 88g; Sodium: 305mg; Carbohydrates: 183g; Fiber: 21g; Protein: 65g

1. In a large skillet over medium-high heat, melt the oil.

2. Add the onion and bell pepper, and sauté until the onion is translucent, 3 to 5 minutes.

3. Add the chicken and sauté until the chicken is cooked through, about 7 minutes.

4. Add the coconut milk, curry powder, salt, and pepper, mix well, and reduce the heat to medium-low.

5. Warm the precooked rice.

6. Top the rice with a big serving of curry and serve.

STORAGE NOTE: This dish keeps in an air-tight container in the refrigerator for up to 1 week.

RECIPE TIP: This is the perfect dish to cook in advance for multiple lunches and dinners. Upgrade the flavor by adding some Fermented Thai Chile Sauce (page 136) and sliced cabbage to the sauté. Replace the chicken with more veggies, such as cauliflower, and you have a meat-less main.

Chocolate Chocolate Brownies, page 127

Chapter Nine

Snacks and Desserts

Sweet Potato Crostini with Feta and Pomegranate Seeds

Serves 1

PREP TIME: 5 MINUTES / COOK TIME: 15 MINUTES

You will be the talk of the town if you show up with these crostini to a dinner party. Get creative with your toppings. This is my gut-healthy favorite.

1 large garnet yam or sweet potato

1 tablespoon avocado oil

Sea salt

Freshly ground black pepper

¼ cup feta cheese (herbed, optional)

½ cup pomegranate seeds

PER SERVING Calories: 371; Total Fat: 23g;
Saturated Fat: 7g; Sodium: 726mg; Carbohydrates: 28g;
Fiber: 5g; Protein: 8g

1. Preheat the oven to 400°F.

2. Slice the yam into thin slices, leaving the skin on. In a large bowl, toss the yam slices with the oil and some salt and pepper.

3. Line a rimmed baking sheet with parchment paper, and spread the yam slices in a single layer without crowding.

4. Roast for 10 to 15 minutes, or until they begin to brown lightly and you can stick a knife into the yam easily but it does not fall apart.

5. Spread the feta cheese onto the crostini, sprinkle with the pomegranate seeds, and serve.

STORAGE NOTE: Make these ahead and refrigerate. Warm them up in the oven, then add toppings right before serving.

RECIPE TIP: You can enjoy these with any of the dips and sauces in this book. Try the Tzatziki (page 139), Homemade Vegan Mayo (page 133), or Guacamole with Pink Cabbage Kimchi (page 119).

Parmesan Zucchini Chips with Tzatziki Dip

Serves 2

PREP TIME: 10 MINUTES / COOK TIME: 20 MINUTES, PLUS 10 MINUTES TO COOL

Parmesan cheese is aged an average of two years. This is significant when it comes to gut health, because the naturally occurring enzymes in certain cheeses help break down the parts of the cheese that are hard to digest, making it easy on the gut. Always opt for older, harder cheeses in small quantities when it comes to gut health.

2 large zucchini, washed, ends cut off

1 teaspoon avocado oil

1 teaspoon garlic salt

Sea salt

Freshly ground black pepper

½ cup Parmesan cheese (easiest if grated and not shredded)

½ cup Tzatziki (page 139)

PER SERVING Calories: 205; Total Fat: 11g; Saturated Fat: 5g; Sodium: 456mg; Carbohydrates: 15g; Fiber: 4g; Protein: 17g

1. Preheat the oven to 450°F.

2. Line a rimmed baking sheet with parchment paper.

3. Slice the zucchini into ¼-inch-thick discs using a knife or a mandoline.

4. Arrange the zucchini in a single layer on the prepared baking sheet. Do not crowd or they will steam and become soggy.

5. Drizzle the zucchini with the oil.

6. Sprinkle with the garlic salt, sea salt, and pepper.

Continued

Parmesan Zucchini Chips with Tzatziki Dip *Continued*

7. Using a small spoon or your fingers, sprinkle each disc with Parmesan.

8. Bake for 10 minutes, then rotate the pan and bake for 10 more minutes, or until the Parmesan begins to brown.

9. Let cool and crisp away from heat for 10 minutes before serving.

10. Serve with a side of Tzatziki.

STORAGE NOTE: This dish is best eaten immediately. If necessary, store in an airtight container in the refrigerator for up to 24 hours.

RECIPE TIP: Precut and refrigerate these zucchini, uncooked, in an airtight container to have ready for addition to a quick meal after work.

Guacamole with Pink Cabbage Kimchi

Serves 4

PREP TIME: 15 MINUTES

When I was first introduced to fermented foods, I'll be honest, I was a little nervous. When I was introduced to this guacamole combo, however, I was addicted. I call this the gateway ferment. Now I only make guac with fermented veggies, and everyone loves them.

4 ripe avocados, peeled and pitted

1 cup Pink Cabbage and Garlic Fermented Kimchi (page 62)

½ cup fresh cilantro, finely chopped

1 tablespoon freshly squeezed lime juice

1 teaspoon sriracha or Fermented Thai Chile Sauce (page 136, optional)

1 garlic clove, minced finely

PER SERVING Calories: 301; Total Fat: 27g; Saturated Fat: 4g; Sodium: 215mg; Carbohydrates: 18g; Fiber: 12g; Protein: 4g

1. In a large bowl, mash the avocados until creamy.

2. If the pink cabbage in the Pink Cabbage and Garlic Fermented Kimchi is cut into long strings, chop it into smaller pieces. Add it to the bowl with the avocado.

3. Add the cilantro, lime juice, sriracha, and garlic, and mix to combine well.

STORAGE NOTE: This dish is best eaten immediately, as avocado browns quickly when exposed to air.

RECIPE TIP: Once you get the hang of fermenting vegetables, you will likely enjoy adding them to guacamole and topping your tacos and fajita bowls with this delicious, gut-friendly condiment.

Stewed Green Apples with Cinnamon

Serves 4

PREP TIME: 5 MINUTES / COOK TIME: 6 HOURS

This dish saved me during my phase of giving up sugar totally. I craved sweetness and desserts so badly, and this warm comforting goodness was always there for me. When everyone else was pigging out on ice cream and cake, I was in heaven with this delight. Guilt-free!

3 pounds Granny Smith apples, peeled and cored

2 tablespoons high-quality butter (optional)

1 teaspoon ground cinnamon

½ teaspoon ground nutmeg

PER SERVING Calories: 90; Total Fat: 0g; Saturated Fat: 0g; Sodium: 2mg; Carbohydrates: 24g; Fiber: 4g; Protein: 1g

1. In the slow cooker, combine the apples, butter (if using), cinnamon, and nutmeg, and set on low for 6 hours, giving it a good stir a few times during cooking.

2. Serve hot.

STORAGE NOTE: This treat is best eaten immediately.

RECIPE TIP: You can also bake these ingredients in a large, covered pot over the stove on medium heat for 25 minutes, stirring often until the apples are tender.

Gluten-Free Granola Clusters

Makes 1 quart

PREP TIME: 20 MINUTES / COOK TIME: 30 MINUTES

This granola doesn't last one day in my house. In addition to being delicious on its own, it's also a great topper over a cup of plain Greek yogurt, kefir, rice cereal, or oats.

1 cup whole raw almonds, divided

1½ cups gluten-free rolled oats

⅔ cup dried Turkish apricots or golden berries

⅓ cup shredded unsweetened coconut

¼ teaspoon sea salt

¼ cup plus 2 tablespoons pure maple syrup

¼ cup coconut oil, melted

PER SERVING (½ cup) Calories: 246; Total Fat: 15g; Saturated Fat: 11g; Sodium: 65mg; Carbohydrates: 26g; Fiber: 4g; Protein: 3g

1. Preheat the oven to 275°F.

2. Line a rimmed baking sheet with parchment paper.

3. Pulse ½ cup of almonds in a food processor or blender until reduced to gritty almond meal. Set aside in a large bowl.

4. Chop the remaining ½ cup of almonds in a food processor or blender until they are large, chunky pieces. Set aside in the bowl with the almond meal.

5. Add the oats, apricots, shredded coconut, and salt to the bowl, and mix well to combine.

6. Add the maple syrup and melted oil to the mixture, using a big spoon or spatula to stir well.

Continued

Gluten-Free Granola Clusters *Continued*

7. Using a small piece of torn parchment, evenly press the mixture into the baking sheet.

8. Roast the granola for 15 minutes, then rotate the pan and roast for 15 more minutes, or until the granola begins to toast lightly and the edges brown.

9. Allow the granola to cool completely before breaking it up into clusters so it doesn't crumble and serve.

STORAGE NOTE: Store in an airtight glass jar in a cool, dark place.

RECIPE TIP: You can really boost the flavor in this recipe by adding sunflower seeds, pumpkin seeds, ground cinnamon, and vanilla extract. Soaking nuts and seeds prior to using is a great way to make them easier to digest, but make sure you have dried them thoroughly before using in the recipe.

Cinnamon Golden Berry Oatmeal Muffins

Makes 6

PREP TIME: 5 MINUTES / COOK TIME: 25 MINUTES

Many people are addicted to the convenience of protein bars; this is my solution. I put these in little baggies and pop them into my purse or in my husband's lunch as dessert or even as a breakfast on the go.

Coconut oil, for greasing the pan

2 cups gluten-free oats

2 overly ripe bananas

½ cup golden berries

⅓ cup almond butter

1 tablespoon ground cinnamon

PER SERVING Calories: 190; Total Fat: 10g; Saturated Fat: 1g; Sodium: 62mg; Carbohydrates: 25g; Fiber: 5g; Protein: 5g

1. Preheat the oven to 350°F. Line 6 cups of a muffin tin with muffin liners or grease.

2. In a large bowl, mix to combine the oats, bananas, berries, almond butter, and cinnamon.

3. Scoop the mixture into the prepared muffin cups.

4. Bake for 20 to 25 minutes, until a toothpick inserted into the center comes out clean, and serve.

STORAGE NOTE: These muffins keep well in an airtight container in the refrigerator for up to 1 week.

RECIPE TIP: You can certainly use raisins here; however, raisins are quite high on the glycemic index since they are much sweeter. Golden berries are more tart and less sugary. Better for digestion, too.

Coconut Macaroon Bombs

Makes 12

PREP TIME: 15 MINUTES / COOK TIME: 20 MINUTES

These little bombs are a guilty pleasure indeed. Easy on the gut and not too sweet, they make the perfect after-dinner treat or midday snack.

1 tablespoon coconut oil

1 tablespoon raw honey

1 tablespoon vanilla extract

1½ cups unsweetened shredded coconut, divided

¼ cup almond flour

3 free-range egg whites

PER SERVING (1 cookie) Calories: 104; Total Fat: 9g; Saturated Fat: 8g; Sodium: 13mg; Carbohydrates: 5g; Fiber: 2g; Protein: 2g

1. Preheat the oven to 350°F.

2. Line a rimmed baking sheet with parchment paper. Set aside.

3. In a small saucepan over medium heat, melt the coconut oil, honey, and vanilla. Set aside.

4. On the baking sheet, lightly toast 1 cup of shredded coconut, about 8 minutes; keep an eye on it so it doesn't burn.

5. In a large bowl, combine the almond flour with the melted oil mixture, and mix well.

6. In a small bowl, whisk the egg whites for 2 minutes, then pour the egg whites and toasted coconut into the almond flour mixture, and mix just enough to combine well.

7. Using a tablespoon, make round scoops out of the coconut dough and roll them in the remaining ½ cup of shredded coconut. Place on the baking sheet.

8. Bake for 8 to 10 minutes, or until the tops are browning, and serve.

STORAGE NOTE: These keep refrigerated for 1 week, and they freeze well, too.

Chocolate-Covered Banana and Nut Butter Bites

Serves 1

PREP TIME: 15 MINUTES / COOK TIME: LESS THAN 5 MINUTES, PLUS 3 HOURS TO FREEZE

This is the messiest, most delicious job you'll ever do. I suggest employing the help of any children in your vicinity. And get ready to lick your fingers, a lot!

3 bananas (not overly ripe)

¼ cup almond butter

1 (9-ounce) package vegan dark chocolate chips (I prefer Lily's brand)

2 teaspoons coconut oil

PER SERVING Calories: 794; Total Fat: 48g; Saturated Fat: 12g; Sodium: 285mg; Carbohydrates: 95g; Fiber: 12g; Protein: 13g

1. Peel the bananas and slice into ¼-inch discs. Place them on a baking sheet lined with parchment.

2. If your nut butter isn't easily spreadable, warm it for up to a minute in the microwave.

3. Using the banana discs and nut butter, make banana sandwiches by spreading enough nut butter so two banana discs stick together, and let the nut butter squish over the sides a bit.

4. Freeze the sandwiches for at least 2 hours, or until they are noticeably firm and frozen.

5. In a microwave-safe bowl, microwave the chocolate and oil on low for 30-second increments, until melted, keeping a close eye and stirring between intervals. The chocolate should be shimmery and very smooth, with no lumps.

Continued

Chocolate-Covered Banana and Nut Butter Bites *Continued*

6. Place another sheet of parchment on another baking sheet, small cutting board, or large plate.

7. Carefully place the hot chocolate bowl near a work space. Take 3 or 4 banana bites from the freezer at a time (so they don't melt), and using 2 spoons or forks, dip the bite thoroughly in the chocolate, then place on the fresh parchment.

8. Freeze the chocolate-covered bites again until firm and frozen, about an hour, and serve.

STORAGE NOTE: You can freeze these treats in an airtight container. They will keep, refrigerated, for 2 weeks.

RECIPE TIP: I choose Lily's chocolate chips because they are made with stevia and totally sugar-free. They can be purchased online. If you are not a fan of using a microwave, you can use a double boiler method:

1. Add 1 inch of water to a medium saucepan or pot and bring to a simmer on the stovetop.

2. Place a heat-safe bowl on top of the pot so the bottom of the bowl is not touching the water.

3. Add the chips to the bowl, and stir occasionally with a spatula until smooth and melted.

4. Remove the bowl from the heat and use your chocolate.

Chocolate Chocolate Brownies

Makes 8

PREP TIME: 30 MINUTES, PLUS 1 HOUR TO FREEZE

Who doesn't like chocolate? This will give you the fix you are looking for. Keep in mind, this is a very high–glycemic–index food. They are most certainly a treat and something great to take to a party.

12 Medjool dates, pitted and soaked for 30 minutes

1 ripe avocado, peeled and pitted

1 cup cashews, soaked for 30 minutes

½ cup unsweetened cocoa powder

1 tablespoon coconut oil

¼ teaspoon sea salt

1 cup Lily's dark chocolate chips, divided

PER SERVING Calories: 315; Total Fat: 23g; Saturated Fat: 10g; Sodium: 74mg; Carbohydrates: 32g; Fiber: 6g; Protein: 4g

1. In a food processor or blender, process the dates, avocado, cashews, cocoa powder, coconut oil, and salt until evenly smooth.

2. Spoon the dough into a large bowl, and fold in most of the chocolate chips until thoroughly dispersed, saving some to sprinkle on top.

3. Press the dough into an 8-inch-by-8-inch baking pan lined with parchment paper, and sprinkle with the remaining chips.

4. Chill in the freezer for 1 hour and serve.

STORAGE NOTE: You can freeze these, or keep them in an airtight container in the refrigerator for up to 2 weeks.

RECIPE TIP: I find that cashews work best with this recipe, and soaking them adds a smoother quality to the finished product.

Homemade Ketchup, page 132

Chapter Ten

Kitchen Staples

Toasted Pumpkin Seeds with Sea Salt

Makes 1 cup

PREP TIME: 1 MINUTE / COOK TIME: 5 MINUTES

So many people ask me why I don't toast nuts and seeds in the oven, and I always respond, "too easy to burn." Lord knows I have ruined my share of oven-toasted nuts and seeds. My solution? Skillet-toasted nuts and seeds. Make a habit out of this simple recipe; you might just use it every day.

½ teaspoon avocado oil

1 cup raw, shelled pumpkin seeds (or pepitas)

Sea salt

PER SERVING (¼ cup) Calories: 50; Total Fat: 4g; Saturated Fat: 1g; Sodium: 60mg; Carbohydrates: 1g; Fiber: 1g; Protein: 2g

1. In a large skillet over medium-high heat, heat the oil until shimmering.

2. Pour the pumpkin seeds into the skillet, and stir occasionally with a spoon.

3. Don't leave the stove; stay put. They will start to snap and brown lightly; continue to stir.

4. Once most of them have snapped and are browned, after 3 to 6 minutes, remove them from the heat.

5. Pour them onto a paper towel and sprinkle with salt.

6. Let them fully cool before storing in a jar with a tight-fitting lid.

STORAGE NOTE: Toasted nuts and seeds can be stored in a cool, dark place for up to 2 weeks. Once you roast nuts and seeds, they go bad much quicker.

RECIPE TIP: Use this method of lightly toasting for all seeds and nuts. Bigger nuts like Brazil nuts should be chopped into small pieces before toasting. I love to also add ¼ teaspoon of various spices; these add such a kick to salads, oats, and just about anything. Try one of these: curry powder, ground cinnamon, ground cayenne, tamari, toasted sesame oil, garlic powder, or chipotle powder.

Homemade Ketchup

Makes 1 quart

PREP TIME: 5 MINUTES, PLUS 2 DAYS TO FERMENT

This is a staple in any household, but as with all store-bought condiments, bottled ketchup contains a lot of preservatives, sugars, and stabilizers, which are all terrible for the gut. This recipe for beneficial ketchup shows you how to make whey in the process!

¼ cup full-fat plain yogurt

3 cups canned tomato paste

3 garlic cloves, peeled and crushed

½ cup grade B maple syrup

1 teaspoon sea salt

2 tablespoons unpasteurized apple
  cider vinegar

PER SERVING (2 tablespoons) Calories: 35;
Total Fat: 0g; Saturated Fat: 0g; Sodium: 84mg;
Carbohydrates: 8g; Fiber: 1g; Protein: 1g

1. Make the whey by pouring the yogurt through a strainer lined with cheesecloth over a bowl. The curds will stay in the cheesecloth, and the liquid (whey) will come through. Retain the liquid in the bowl, and discard the curds.

2. Add the tomato paste, garlic, maple syrup, salt, and vinegar to the bowl, and mix until well blended.

3. Pour the ketchup into a quart-size, widemouthed jar. The top of the ketchup should be at least 1 inch below the top of the jar. Cover.

4. Leave at 72°F to 75°F for 2 days before refrigerating.

STORAGE NOTE: This ketchup keeps in the refrigerator for at least 1 month in an airtight jar.

RECIPE TIP: Grade B maple syrup is common; even Trader Joe's has its own brand. There are a few other ingredients in the famous original *Nourishing Traditions* cookbook ketchup recipe. This would include ½ cup fish sauce and ¼ teaspoon cayenne pepper. You can also add ground turmeric and ground cumin.

Homemade Vegan Mayo

Makes 2 cups

PREP TIME: 20 MINUTES, PLUS OVERNIGHT TO SOAK

A healthier mayonnaise is better for your digestion, and you'll never know the difference. It's creamy and tangy and can be used in the same ways as regular mayo: on a sandwich, in a tuna salad, as a raw veggie dip at parties, in deviled eggs, or in my favorite, Baby Red Potato Salad (page 72).

½ cup raw almonds

¼ teaspoon garlic powder

¾ teaspoon sea salt

½ to ¾ cup water

1 cup flax oil

3 tablespoons freshly squeezed lemon juice

½ teaspoon unpasteurized apple cider vinegar

PER SERVING (1 tablespoon) Calories: 69; Total Fat: 8g; Saturated Fat: 1g; Sodium: 55mg; Carbohydrates: 0g; Fiber: 0g; Protein: 0g

1. In a medium bowl, soak the almonds overnight.

2. In the morning, slip the skins off the almonds and pat the almonds with a paper towel until dry.

3. In a blender or food processor, pulse the almonds into a fine powder.

4. Add the garlic powder and salt, and blend well.

5. Add the water, and blend until creamy.

6. Keep your blender or processor on low, and very slowly drizzle in the oil until the mixture thickens.

7. Add the lemon juice and vinegar, and blend for 1 minute, until the mixture has a mayonnaise consistency.

STORAGE NOTE: This recipe will keep, refrigerated, in an airtight container for 2 weeks.

RECIPE TIP: If you have a tree nut allergy, replace the almonds with 2 egg yolks. Also, note that the slower you drizzle the oil into the mixture, the creamier the texture becomes.

Homemade Coconut Milk

Makes 2 cups

PREP TIME: 15 MINUTES

Store-bought coconut milk is loaded with unwanted ingredients like stabilizers, sugars, and flavorings. Make your own, and save money and be kind to your gut. See the recipe tip for ideas for its use.

1 cup shredded coconut meat, dry or fresh
(if dry, soak in warm water for 15 minutes, then strain)

2 cups water

PER SERVING (½ cup) Calories: 23; Total Fat: 2g; Saturated Fat: 2g; Sodium: 0mg; Carbohydrates: 1g; Fiber: 1g; Protein: 0g

1. In a high-powered blender, liquefy the coconut and water for 2 to 3 minutes.

2. Using cheesecloth, a nut milk bag, or a large, fine mesh strainer, pour the contents into the cloth or bag and strain well.

STORAGE NOTE: Store in an airtight jar in the refrigerator for 1 week.

RECIPE TIP: Coconut milk can be used to make curries, soups, smoothies, or a warm beverage by adding ground cinnamon, ground cardamom, ground nutmeg, and cacao.

Vinaigrette Four Ways

Makes 1 cup

PREP TIME: 10 MINUTES

Vinaigrettes are so versatile. Not just limited to salads, they can be served with fruit, quinoa, or rice and sprinkled on avocado toast, to name a few ways to enjoy them.

Basic Vinaigrette

¾ cup olive oil

¼ cup unpasteurized apple cider vinegar

1 tablespoon Dijon mustard

1 garlic clove, finely minced

Sea salt

Freshly ground black pepper

Lemon Herb Vinaigrette

Basic Vinaigrette ingredients plus:

1 tablespoon freshly squeezed lemon juice

½ cup finely chopped fresh dill

Raspberry Vinaigrette

Basic Vinaigrette ingredients plus:

¼ cup fresh raspberries, puréed in a blender

Asian Vinaigrette

Basic Vinaigrette ingredients plus:

2 tablespoons toasted sesame oil

For each recipe, combine all the ingredients in a small jar with a lid, seal, and shake vigorously.

STORAGE NOTE: All these dressings can keep, refrigerated, for up to 2 weeks.

RECIPE TIP: I recommend always having multiple jars of vinaigrette on hand in your fridge.

PER SERVING (**2 tablespoons**) Calories: 165; Total Fat: 19g; Saturated Fat: 3g; Sodium: 55mg; Carbohydrates: 0g; Fiber: 0g; Protein: 0g

Fermented Thai Chile Sauce

Makes ½ cup

PREP TIME: 10 MINUTES, PLUS 3 DAYS TO FERMENT

According to a study done by the University of Connecticut, hot peppers may help reduce inflammation in some people with gastric issues. Of course, some sensitive tummies can't do spicy, but for those of you who can, this is a vitamin-rich, anti-inflammatory fermented powerhouse.

1 cup Thai chiles, stemmed and
    seeded removed

4 garlic cloves

¼ cup fresh parsley

1 teaspoon sea salt

PER SERVING (1 tablespoon) Calories: 6; Total Fat: 0g; Saturated Fat: 0g; Sodium: 265mg; Carbohydrates: 1g; Fiber: 0g; Protein: 0g

1. In a blender (see recipe tip) or food processor, blend to combine the chiles, garlic, parsley, and salt until you have a chunky purée.

2. Scoop out the mixture into a small jar with a tight-fitting lid.

3. Store at 72°F to 75°F for 3 days. Be careful not to let the temperature drop below 72°F or the fermentation may stop. Use a heating pad if necessary.

4. After 3 days, refrigerate and use sparingly with your food. Depending on the chiles you picked, this sauce can be incredibly hot, and the fermentation process can make them even hotter. Use caution when eating.

STORAGE NOTE: This chili sauce can be kept refrigerated for a very long time. Unopened it will keep for literally years! Once you open it, it can be stored for up to 1 month.

Healthy BBQ Sauce

Makes 1 cup

PREP TIME: 5 MINUTES / COOK TIME: 10 MINUTES

Store-bought barbecue sauce is loaded with artificial flavorings, preservatives, and sugars that wreak havoc on the gut. This sauce is easy and delicious—and it's clean, meaning it doesn't contain all the mysterious ingredients in store-bought versions. Your gut will thank you.

1 (6-ounce) can tomato paste

½ cup unpasteurized apple cider vinegar

1 teaspoon chili powder, or more or less

½ teaspoon garlic powder

1 teaspoon Dijon mustard

Sea salt

PER SERVING (2 tablespoons) Calories: 23;
Total Fat: 0g; Saturated Fat: 0g; Sodium: 63mg;
Carbohydrates: 5g; Fiber: 1g; Protein: 1g

1. In a small saucepan over medium-low heat, combine the tomato paste, vinegar, chili powder, garlic powder, and mustard, and season with salt. Bring to a very low simmer.

2. Stir and simmer until the sauce thickens to your liking, about 10 minutes.

STORAGE NOTE: You can store this sauce for up to 2 weeks in an airtight container in the refrigerator.

RECIPE TIP: One of the best parts of regular barbecue sauce is the sweetness. If you miss that in this sauce, you can add 1 teaspoon of raw honey.

Vegan Pesto

Makes 1 cup

PREP TIME: 20 MINUTES

On my first day of culinary school, my chef said, "Get good at making sauces: they make everything good." She was so right. I have my go-tos, and this vegan pesto is one of them. I use it as a dip, a pasta sauce, a sandwich spread, on breakfast toast with avocado, as a sauce for fish or meat, and as a salad dressing.

2 cups fresh basil or spinach

2 garlic cloves, roughly chopped

½ cup walnuts, pine nuts, or soaked cashews

Sea salt

Freshly ground black pepper

½ cup olive oil

1 tablespoon freshly squeezed lemon juice

PER SERVING (1 tablespoon) Calories: 80; Total Fat: 9g; Saturated Fat: 1g; Sodium: 4mg; Carbohydrates: 1g; Fiber: 0g; Protein: 1g

In a food processor or high-powered blender on medium speed, pulse or blend the basil, garlic, and nuts, and season with salt and pepper. Then lower the speed and slowly add the olive oil and lemon juice, blending until smooth but still a little coarse.

STORAGE NOTE: This stores well in an airtight container in the refrigerator for 1 week. You can also freeze it in individual containers and use as needed.

RECIPE TIP: A food processor works best for this job, but a high-powered blender is fine, too. Freeze pesto in ice cube trays and empty into a pot to melt down with a bit of butter. Add to the top of some fish to flavor it up, or pop some onto a baked yam. You can produce this effect with many herbs—great flavor is just an ice cube away.

Tzatziki

This creamy Greek dip or sauce is known for its pairing with meats, but it can be used with a wide range of dishes from meats to vegetables, and even as a salad dressing.

1 cup grated English cucumber

2 cups Greek yogurt

4 garlic cloves, finely minced

2 tablespoons freshly squeezed lemon juice

1 tablespoon olive oil

1 tablespoon fresh dill, chopped small

Sea salt

Freshly ground black pepper

PER SERVING (¼ cup) Calories: 39; Total Fat: 2g;
Saturated Fat: 1g; Sodium: 16mg; Carbohydrates: 2g;
Fiber: 0g; Protein: 4g

1. Squeeze as much fluid out of the grated cucumber as you can.

2. In a medium bowl, combine the cucumber with the yogurt, garlic, lemon juice, olive oil, and dill, and season with salt and pepper. Mix well.

STORAGE NOTE: Tzatziki can be refrigerated in an airtight container for 1 week.

RECIPE TIP: You can replace the Greek yogurt with plain coconut yogurt or 2 cups of cashews (soaked overnight). Combine all the ingredients in a blender and blend until smooth.

Homemade Kefir, page 58

Measurement Conversions

Volume Equivalents (Liquid)

US STANDARD	US STANDARD (ounces)	METRIC (approximate)
2 tablespoons	1 fl. oz.	30 mL
¼ cup	2 fl. oz.	60 mL
½ cup	4 fl. oz.	120 mL
1 cup	8 fl. oz.	240 mL
1½ cups	12 fl. oz.	355 mL
2 cups or 1 pint	16 fl. oz.	475 mL
4 cups or 1 quart	32 fl. oz.	1 L
1 gallon	128 fl. oz.	4 L

Oven Temperatures

FAHRENHEIT (F)	CELSIUS (C) (approximate)
250°F	120°C
300°F	150°C
325°F	165°C
350°F	180°C
375°F	190°C
400°F	200°C
425°F	220°C
450°F	230°C

Volume Equivalents (Dry)

US STANDARD	METRIC (approximate)
⅛ teaspoon	0.5 mL
¼ teaspoon	1 mL
½ teaspoon	2 mL
¾ teaspoon	4 mL
1 teaspoon	5 mL
1 tablespoon	15 mL
¼ cup	59 mL
⅓ cup	79 mL
½ cup	118 mL
⅔ cup	156 mL
¾ cup	177 mL
1 cup	235 mL
2 cups or 1 pint	475 mL
3 cups	700 mL
4 cups or 1 quart	1 L

Weight Equivalents

US STANDARD	METRIC (approximate)
½ ounce	15 g
1 ounce	30 g
2 ounces	60 g
4 ounces	115 g
8 ounces	225 g
12 ounces	340 g
16 ounces or 1 pound	455 g

The Dirty Dozen™ and the Clean Fifteen™

A nonprofit environmental watchdog organization called Environmental Working Group (EWG) looks at data supplied by the U.S. Department of Agriculture (USDA) and the Food and Drug Administration (FDA) about pesticide residues. Each year it compiles a list of the best and worst pesticide loads found in commercial crops. You can use these lists to decide which fruits and vegetables to buy organic to minimize your exposure to pesticides and which produce is considered safe enough to buy conventionally. This does not mean they are pesticide-free, though, so wash these fruits and vegetables thoroughly.

THE DIRTY DOZEN™*

1. apples
2. celery
3. cherries
4. grapes
5. nectarines
6. peaches
7. pears
8. potatoes
9. spinach
10. strawberries
11. sweet bell peppers
12. tomatoes

* Additionally, nearly three-quarters of **hot pepper** samples contained pesticide residues

THE CLEAN FIFTEEN™

1. asparagus
2. avocados
3. broccoli
4. cabbages
5. cantaloupes
6. cauliflower
7. eggplants
8. honeydew melons
9. kiwis
10. mangoes
11. onions
12. papayas
13. pineapples
14. sweet corn
15. sweet peas (frozen)

Resources

ONLINE

If you find that you'd like to continue *Gut Health Diet for Beginners* but need a little more guidance, you can join the Total Gut Makeover by visiting HealthyGutGirl.com, where I will guide you through a six-week program of handheld gut help.

I also host the podcast called Stuff Your Doctor Should Know. It offers a diverse array of nutritional information, as well as guests who are world-renowned authors, scientists, and regular people with incredible journeys. You can find the podcast on iTunes, Spotify, or by visiting HealthyGutGirl.com

For high-quality nutraceutical supplements without any toxic preservatives or excipients, sign up for your own free account with Premier Research Labs by visiting HealthyGutGirl.com and clicking on the Shop page.

BEDROKCommunity.org
Autism Recovery and BEDROK community. Join for more information on recovery of autism in children through diet.

MonashFODMAP.com/i-have-ibs/get-the-app
Download the low FODMAP diet app for IBS support and the largest FODMAP food database available.

WestonAPrice.org
The Weston A. Price Foundation is an excellent source for accurate information on nutrition and health, always aiming to provide the scientific validation of traditional food.

DedicatedWellness.com
Chiropractic in relation to gut health, with Dr. Charles Martone.

BOOKS

The Body Ecology Diet, by Donna Gates, M.Ed., ABAAHP
This international best-selling author is an advanced fellow with the American Academy of Anti-Aging Medicine and is on a mission to change the way the world eats.

Gut and Psychology Syndrome, by Dr. Natasha Campbell-McBride
The author holds a degree in medicine and postgraduate degrees in nutrition. In her clinic in Cambridge, she specializes in nutrition for children and adults with behavioral and learning disabilities as well as adults with digestive and immune system disorders.

Nourishing Traditions, by Sally Fallon
Author Sally Fallon is founder of The Weston A. Price Foundation. The book is filled with "old-school" recipes and encourages us to look to our ancestors' recipes and ways for healing and balance now.

The Human Superorganism, by Rodney Dietert
This book explores how the microbiome is revolutionizing the pursuit of a healthy life.

I Contain Multitudes, by Ed Yong
The author explores the microbes within us and takes a grander view of life.

References

Calafat, Antonia M. et al. "Polyfluoroalkyl Chemicals in the U.S. Population: Data from the National Health and Nutrition Examination Survey (NHANES) 2003–2004 and Comparisons with NHANES 1999–2000." *Environmental Health Perspectives* 115, no. 11 (2007): 1596–1602. PMC. Web. 30 Aug. 2018.

Clapp, Megan, Nadia Aurora, Lindsey Herrera, Manisha Bhatia, Emily Wilen, and Sarah Wakefield. "Gut Microbiota's Effect on Mental Health: The Gut-Brain Axis." *Clinics and Practice* 7, no. 4 (Sept. 15, 2017). www.ncbi.nlm.nih.gov/pmc/articles/PMC5641835/

DeBaun, Daniel T.. *Radiation Nation: Fallout of Modern Technology.* Icaro Publishing, 2017.

Evrensel, Alper, and Mehmet Emin Ceylan. *The Gut-Brain Axis: The Missing Link in Depression.* Dec. 13, 2015. doi: 10.9758/cpn.2015.13.3.239

Feltman, Rachel. "The Gut's Microbiome Changes Rapidly with Diet." *Scientific American.* December 2013. www.scientificamerican.com/article/the-guts-microbiome-changes-diet/

Fields, Douglas. "Mind Control by Cell Phone," *Scientific American.* May 7, 2008. https://www.scientificamerican.com/article/mind-control-by-cell/

Foster, Jane A., Linda Rinaman, John F. Cryan. "Stress and the Gut-Brain Axis: Regulation by the Microbiome," *Neurobiology of Stress* 7. December 2017. Dec. 7, 2017. www.sciencedirect.com/science/article/pii/S2352289516300509

Harrison, Laird. "Antibiotics Still Overprescribed for Sore Throats, Bronchitis." Oct. 4, 2013. MedScape. www.medscape.com/viewarticle/812109

Johnson, Katerina V.-A. and Philip W. J. Burnet. "Microbiome: Should We Diversify from Diversity?" *Gut Microbes* 7, no. 6 (Oct. 17, 2016): 455–458. www.ncbi.nlm.nih.gov/pmc/articles/PMC5103657/

Quigley, Eamonn M. M. "Gut Bacteria in Health and Disease." *Gastroenterology & Hepatology* 9, no. 9 (2013): 560–569. www.ncbi.nlm.nih.gov/pmc/articles/PMC3983973/#B2

University of Connecticut. "Chili peppers and marijuana calm the gut, study suggests: The active ingredients in both hot peppers and cannabis calm the gut's immune system." ScienceDaily. April 2017. www.sciencedaily.com/releases/2017/04/170424152537.htm

Wipler , J., Z. Čermáková,
    T. Hanzálek, H. Horáková,
    H. Žemličková. Čermáková
    Z, Hanzálek T, Horáková
    H, Žemličková H. "Sharing Bacterial
    Microbiota Between Owners and
    their Pets." *Klin Mikrobiol Infekc
    Lek* 23, no. 2 (June 23, 2017):
    48–57. www.ncbi.nlm.nih.gov
    /pubmed/?term=Wipler%20
    J%5BAuthor%5D&cauthor=
    true&cauthor_uid=28903168

World Health Organization. "Antibiotic
    Resistance - A Threat to Global
    Health Security." Presented at
    World Health Organization, Geneva,
    Switzerland, May 2013. www.who
    .int/drugresistance/activities
    /wha66_side_event/en/

Yong, Ed. *I Contain Multitudes*. Ecco
    Publishing, 2016.

Recipe Index

Index

Acknowledgments

The three mentors who changed my life and were guiding lights: Donna Gates, Dr. Robert Marshall, and my husband, Dr. Charles Martone. They helped give my life true meaning. I spent years not contributing to humanity in any real, useful way, and it wasn't until I did that I understood exactly what it meant to give of myself in that way. And so here I stand on the shoulders of giants, spreading a beautiful word that is a mixture of all their wisdom. Forever grateful. And thank you to my friends and family who are my biggest fans. And to the team at Callisto for knowing I was fully capable and trusting my experience.

About the Author

Kitty Martone is a holistic health practitioner, a master herbalist, and a chef. With this background, and using her own health challenges as stepping stones to further her education, she has learned and will continue to learn about the ever-changing world of the human microbiome and how it is the conductor of our health and wellness. During her darkest days with her own health challenges, she prayed for guidance—the kind of guidance that she can now offer through her podcast, Stuff Your Doctor Should Know, and through her thriving practice and writing. It's a dream come true for her to be able to connect you with alternative methods of healing, scientists on the cutting edge of their field of work, and so much more. You can join the Healthy Gut Girl community here:

Facebook: Healthy Gut Girl

Facebook: Estrogen Dominance Support Group

iTunes: Stuff Your Doctor Should Know

Instagram: @HealthyGutGirl_

CPSIA information can be obtained
at www.ICGtesting.com
Printed in the USA
LVHW07s2346251018
594834LV00003B/5/P